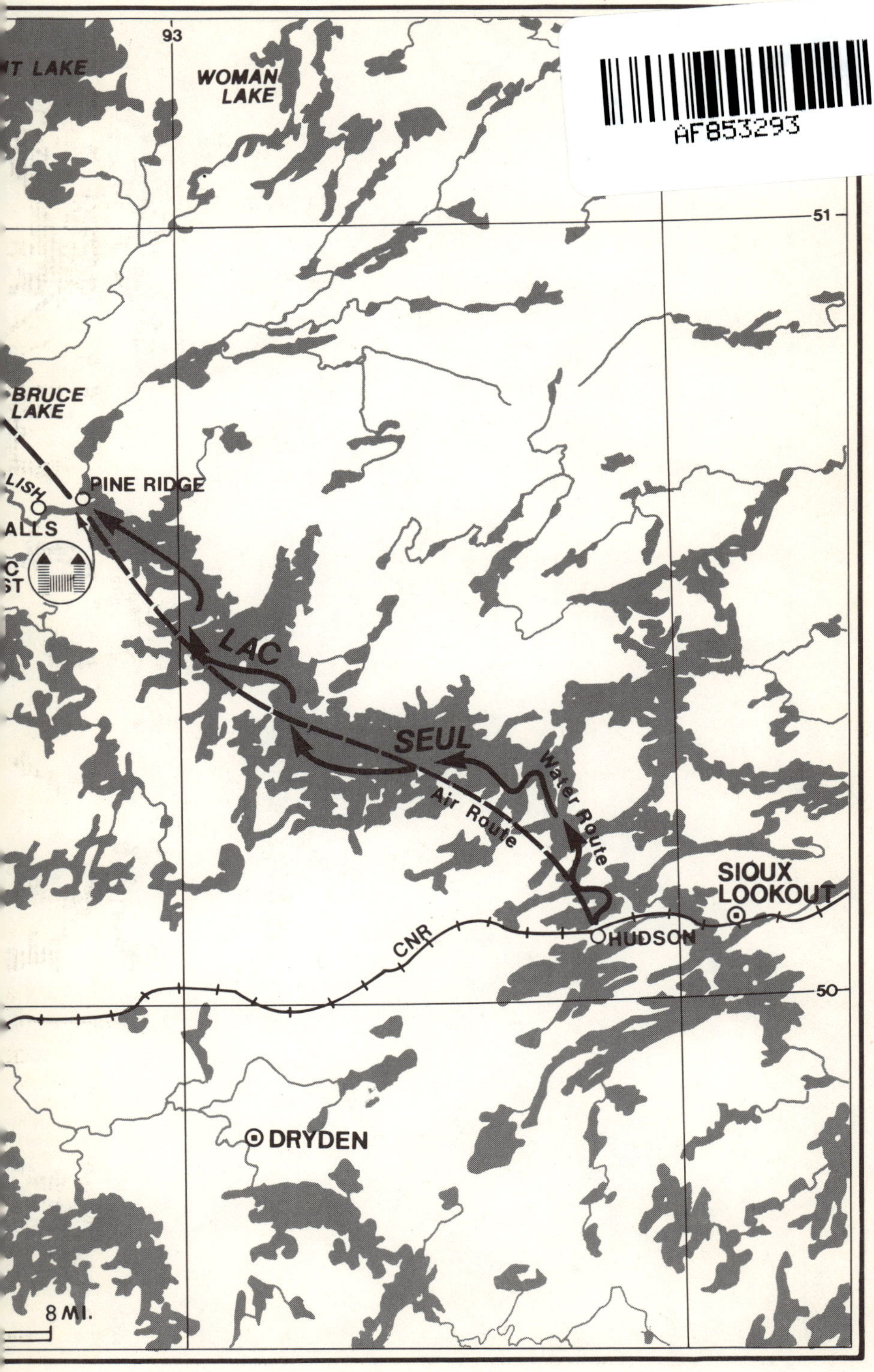

93
T LAKE
WOMAN LAKE
51
BRUCE LAKE
PINE RIDGE
LISH
ALLS
C
ST
LAC
SEUL
Water Route
Air Route
SIOUX LOOKOUT
HUDSON
CNR
50
DRYDEN
8 MI.

July 18, 1987

North for Gold

The Red Lake Gold Rush of 1926

Ruth Weber Russell

Ruth Russell

North Waterloo Academic Press 1987

ISBN 0-921-075-01-04

Canadian Cataloguing in Publication Data

Russell, Ruth Weber.
North for Gold

Includes index.
ISBN 0-921-075-01-04

1. Gold mines and mining - Ontario - Red Lake Region - History.
2. Red Lake Region (Ont.) - Gold discoveries.
3. Red Lake Region (Ont.) - History. I. Title.

FC3099.R43R88 1987 971.3'112 C87-094537-8
F1059.5.R43R88 1987

North Waterloo Academic Press
370 University Avenue East
Waterloo, Ontario N2K 2B5
Printed and bound in Canada by
John Deyell Company, Lindsay, Ontario
1987

This book is dedicated to the memory of my father,
Andrew Weber.

Biographical Note

Ruth Weber Russell was born and educated in Red Lake, where her father worked from 1940 until the early 1970s for Madsen Red Lake Gold Mines. She graduated from Red Lake District High School in 1963 as the school's first Ontario Scholar and returned to work summers in the Madsen Mine office while a student at the University of Toronto, where she was granted a B.A. and M.A. in English.

She has worked as an editor for Copp Clark in Toronto, as a journalist for *The North Bay Nugget,* and for the last decade has lived with her husband and two daughters in Kitchener, where she is Coordinator of the Waterloo Regional Arts Council, and edits the Arts Council's bi-monthly *Arts Now.*

In addition to freelance articles and journalism, she has published a cookbook, and is co-author of *Lucy Maud Montgomery: A Preliminary Bibliography* (1986).

Acknowledgements

My thanks go to Campbell Red Lake Mines Limited, and in particular to Campbell's Manager, Stewart Reid, for a research grant to me for pursuing this project. I must also acknowledge my debt to Don Parrott's privately-printed local history, *The Red Lake Gold Rush,* first published in 1964. I also want to express my thanks to the Ontario Ministry of Citizenship and Culture (in particular to Michelle Greenwald) for opening its files on Red Lake to me; to the staff at the Ministry of Natural Resources in Red Lake, especially to Steve van Haaften and Kate van Leeuwen; to Ray Riley, of the Ministry of Natural Resources, for lending material and finding answers to a variety of questions; to Harold Farrington and Margaret McDougall; to George Aiken and Kay Tingley, both historians of the Red Lake area; to Joyce Appel, Curator of the Ear Falls museum; to archivists at the Archives of Ontario and at the Hudson's Bay Archives in Winnipeg; to Professor Elizabeth Arthur, at Lakehead University; to Bob and Pat Ballantyne of Waterloo; to Catherine Pepper at the Canadian Red Cross; to the librarians at the University of Waterloo library, for their unfailing help; to my husband, for critical reading and technical help.

Contents

Introduction

Sixty years ago there was a gold rush towards a spot deep in the inaccessible bush of northwestern Ontario. The rush began with a modest discovery of gold on the shores of Red Lake by two brothers from Haileybury in July of 1925. It took some time for the news to spread and for the rush to gather its own self-propelling momentum, but by early 1926 it was in full swing. Thousands of men rushed to Red Lake by any available means: most came by dogsled, but some came by airplane, by horse-drawn sleigh, by primitive snow machines (until they broke down), by foot and, in the spring, on anything that would float. The gold rush was short-lived, however, and by Christmas of 1926 it was all over. And yet, the real development of Red Lake into the major gold-producing area it later became, had yet to begin. The gold rush of 1926 was the last of its kind in Canada, largely because the technological innovations pioneered during the Red Lake gold rush made such primitive mass movement unnecessary. Yet this rush can be seen as a significant force in the settlement patterns which have changed Canada from vast, uninhabited forest into the modern industrial nation that exists today. The history of the Red Lake gold rush is a fascinating story of adventure, of hope, of disappointment, and of some remarkable luck and persistence. It was an adventure of the common man, the kind of uncomplicated adventure that is no longer possible in our sophisticated society, and it deserves a niche of its own in the records of Canadian history.

Gold has been perceived as an extremely precious commodity since it was first mined, around 3900 B.C. Many of gold's principal uses have not changed for thousands of years: gold as a monetary commodity, gold as coinage, gold as jewellery. (To these have been added, in more recent times, industrial uses for gold: in electronics, where it is valued for its extremely low susceptibility to corrosion and its high conductivity; in medical and dental technology; in the production of nitric acid, of synthetic threads, of solar cells.)

As a highly prized commodity, gold has had its current price monitored for over 800 years. In 1717 Sir Isaac Newton,

Master of the Mint in London, fixed the price of gold at 84 shillings, eleven pence halfpenny per troy ounce. With certain variations during the First World War, this price lasted until 1934. (In 1926, the year of the Red Lake gold rush, this translated into $20.67, Canadian dollars.) In 1934 the U. S. A. fixed the price of gold at $35.00 U.S. per ounce, and it remained close to that figure until March of 1968 when the price was set free to float.

Since that point, the price of gold has fluctuated widely. It hit a new height of $200 an ounce in 1978, another of $300 in 1979, and the unimagined peak of $800 an ounce in 1980. In 1982, however, it sank to $325. Although there is always a market for gold, the actual quantities mined at any particular time depend, to some extent, on the current price. (So, indeed, does the quality of ore mined at any particular time: when the price of gold is low, a high grade of ore is mined; when the price is high, a lower grade is mined, and thus the mine's profit per ton remains relatively constant.)

About 90% of the world's gold production comes from four countries: South Africa, Russia, Canada and the U.S.A., in that order. It is deceiving, however, to think of this as a like group. South Africa alone produces half the world's output of gold, from a group of very deep and very rich mines in a 300-mile arc to the west and south of Johannesburg. Of the world's 20 largest gold mines, 17 are in South Africa. Of a world-wide total production in 1978 of 1,210 metric tonnes, South Africa produced 706.4 tonnes; the U.S.S.R., 250 tonnes; Canada, 54 tonnes (of this, Ontario produced 21.8 tonnes and Campbell Red Lake, 5.7 tonnes), the U.S., 30.2 tonnes.

In spite of the ancient history of gold-mining, the gold rush itself is a relatively modern phenomenon. It was not until the nineteenth century that the hunger for gold inspired hordes of men (and a very few women) to rush to, what were then considered, the ends of the earth in search of this precious commodity. In the last century, great gold rushes occurred in every corner of the world: in Brazil, in Siberia, Australia, South Africa, California, and the north of Canada. In both Australia and South Africa, the discovery of gold and the consequent rush of prospectors to reported sites led to the eventual settlement of vast areas of land and therefore, for better or for worse, laid many foundations for contemporary society in these countries. (The strict racial division of South African society owes much to the large-scale and undemocratising mining companies.)

In a similar way, the gold rushes in North America in the last century were also ultimately responsible for the permanent settlement of previously uninhabited territory. The gold rush to California in 1849 brought over 80,000 people to California in that year, and laid the foundations for, not only a long-lasting mining industry on the west coast, but for the prosperous state of California. Similar results followed the two gold rushes that occurred shortly thereafter, to the north, in British Columbia. At least 25,000 people rushed to the Fraser River in 1858, and a smaller number to the Cariboo four years later. Both British Columbia rushes were considerably more orderly than the great rush of the 49ers to California. (This was owing to the prompt and efficient establishment, by the British, of governmental machinery, and not to any difference in personnel. For most of the men who rushed to the gold finds in B.C. were men who had rushed to those in California years before.) Although most of the "fortune-seekers" drifted on elsewhere after the B.C. rushes were over, the impact of the rushes had transformed a few fur-trading outposts on the Pacific coast into a true colony which was shortly to view itself as part of the future nation of Canada.

The impact of the Klondike gold rush, that began in 1896, was not as clear-cut as those previously mentioned. The event itself was an unprecedented one; some 50,000 men made the long and incredibly difficult trek to the Klondike over hazardous mountain passes, and most travelled back out again. After the gold production had peaked in the year 1900 (at 1,350,000 ounces) the population began to drop, and it dwindled rapidly thereafter. The fact became apparent that, unlike the gold rushes to the south, in British Columbia and California, the Klondike rush had occurred in an area that, because of its climate, its terrain and its isolation, could not support a large population. And yet, there were some long-lasting settlement gains that stemmed from the Klondike rush. These occurred, however, not in the remote Klondike region itself, but in places that had served as supply and transportation despatching centres for people bound for the Klondike: in the state of Washington, in the province of British Columbia, and in western Canada in general.

The next significant gold rush in Canada sprang out of the silver rush to Cobalt, one of the major mining stampedes in history, that peaked in 1905. Gold discoveries in neighbouring areas of northeastern Ontario (Porcupine, Kirkland Lake, Gowganda) followed rapidly as men came in search of silver. Towns developed quickly, and most of these are still in existence. The area, though a good deal closer to urban amenities than was the Klondike at the time of its rush, was still too far away, too cold, and surrounded by land of too little agricultural value, to support a large population. And yet, as the Klondike mining finds gave an important impetus to the southerly towns which acted as their supply depots, so the northeastern Ontario developments gave an enormous financial boost to their major supply depot to the south, Toronto.

The Cobalt/Porcupine mining development was tied to that of Red Lake in many ways. Among these connecting threads is the fact that many of the people who rushed to Red Lake had earlier rushed to Cobalt/Porcupine and, therefore, had established a set of expectations for Red Lake. (This happened, not just in Canada, but throughout the world; a world that, by 1926, was already shrinking through the power of the press.)

The role which the press played in the Red Lake gold rush was a very large one. Newspaper reports of the gold discoveries, there, were responsible for almost all of the "hopefuls" who rushed to Red Lake in the peak months of the stampede. And, similarly, the sudden decline in press coverage, in the latter part of 1926, helped to actively determine when the rush was over.

Well before the Red Lake gold rush in 1926, gold mining had ceased to be a poor man's game. (The Klondike was an anomaly in this pattern, for, in the Klondike, a man with a pick, a shovel and a pan could, and sometimes did, make a fortune. Gold there was "placer gold", lying on stream beds or close to them, easily accessible and separable from its surrounding material by a method as primitive as shaking it in a pan with water.) The enormous gold mining area of the Rand in South

Africa was an immense body of low-grade ore which contained "lode gold". This is gold which has been deposited in lodes (or veins) in the rock and therefore requires large sums of money and elaborate technology to extricate it. The gold at Porcupine, and Red Lake was lode gold as well.

The gold rushes of history contain a central inconsistency. Ideally, the gold rushes are adventures of the common man; they were, by 1926, one of the few enterprises left wherein one man alone, with a few tools and a few dollars, could set out to conquer the world and make his fortune. But, in reality, the gold rushes were enterprises in which the vast majority of participants failed to even recoup their expenses. There were a very few "little men" who were lucky enough to be at the right place at the right time, early in the staking rush, in Rand, in Porcupine, and at Red Lake, and only these lucky few accrued large sums of money.

The gold rush was an international phenomenon, and the gold rush to Red Lake was part of the pattern of gold rushes that occurred around the world over the previous century. But the Red Lake rush can also be seen as a part of other national patterns, Canadian patterns. The patterns of settlement in the vast uninhabited wilderness of Canada, include a number of irregularities which are caused by places like Red Lake. From the time of Champlain, the Canadian frontier of settlement was pushed gradually southwestward until it stopped at the Canadian Shield and the end of the St. Lawrence waterway. The frontier stopped there temporarily, while new settlements emerged from the other end of the country, on the Pacific coast. When the railway joined east and west together, settlement began to spread out from the railway line. The principal function of settlements during this period was to develop agricultural land and, therefore, in the fertile areas like southern Manitoba, settlement followed immediately on the heels of the railway.

By about 1920, the settlement phase of the agricultural pioneer was over. Not only the good, but the not-so-good land had been settled. The future development of Canada was industrial, in manufacturing and resource developments. Both of these new facets of Canadian life began to boom after 1920.

Until this happened, though, in areas conspicuously lacking in arable land (like the long stretch of Canadian Shield country between the head of Lake Superior and the beginning of the prairies) the only settlement activity at all occurred right along the railway line. And for the last hundred years, from the date that this stretch of railway was completed until today, there have been only two causes for settlement north of it: the forest industry and the mining industry. The map of that part of Canada today is still virtually void of names that signify human habitation; the longest highway going north from the Trans-Canada Highway (which follows the route of the CPR line) is Highway 105 to Red Lake, and beyond that is the forest.

From the time of the big push in mineral development at the beginning of this century (of which the gold rush to Red Lake was a direct result) there has been a continuing increase in the importance of Canada's mineral development. However, gold mining did exist before this century in Canada, and not just in the big western rushes. The first gold found in Canada was in 1823, at the mouth of a tributary of the Chaudière River in Quebec. Gold mining began in Nova Scotia in the middle of the last century. The first gold discovery in Ontario was made in 1866 on a farm near Madoc, and in the 1890s a number of gold discoveries were made and mined in the Lake of the Woods area. Gold-mining itself reached a production peak in 1940 which it has not attained since. Developments since World War II in the exploitation of base metals and of oil and gas, however, have continued to make the resource industry a major bulwark of the Canadian economy.

The expansion of mining development in the 1920s seems, in retrospect, just one expression of the mood of optimism and expansionist thinking that was found in Canada as a whole during that decade. (This was quite different from the "depression boom" in gold-mining that occurred in the 1930s, when gold-mining was one of the few bright spots in the economy.)

In the mid-twenties the whole Canadian economy was booming. The substantial grain trade, which until then had been the backbone of the Canadian economy, remained very strong, and to it was added new growth in areas like manufacturing and exploitation of mineral and forest resources. Unemployment was very low, and the wages of the average family had risen moderately against the cost of living. As an article in *Canadian Forum* pronounced, "It is now . . . generally admitted that we are in a period of prosperity. Canada has emerged from the lean years . . . "[1] Because it was a period of economic optimism, it was also one of increased activity on the stock market. The penny mining stocks were particularly attractive to the small investor with a few dollars to invest in his hopes; he knew of the important mining fields that had developed so rapidly in northeastern Ontario, and he had every hope that new finds in the romantically distant Patricia region would prove as rich.

The amount of money invested by the average speculator in the Red Lake stocks floated on the Mining Exchange, however, was far less than the amount spent by the average man who rushed to Red Lake to look for gold himself. It is possible only to estimate what it cost the average gold-seeker to get to Red Lake, spend at least several weeks there, and get back out again, but a conservative estimate of that amount is $500 per person. (This total would be easily reached by adding together the expenses of taking the train from the prospector's point of origin to Hudson, and from there either flying to Red Lake–which was never cheaper than $200 return during the gold rush, and which was twice that at the beginning of the rush–or buying a dog-team and having the expense of feeding the dogs, as well as himself, while at Red Lake.)

The Howey brothers, whose finds started the gold rush, made significant amounts of money, as did a few others who staked their claims in the last months of 1925 and in January of 1926, before the rush began. But after the rush had started, the largest profit that any claim-stakers attained was what the *Manitoba Free Press* called a "comfortable little stake" that a prospector might get when he sold a claim or two. The modest sum that an old-time prospector thus gained he would use, speculated the journalist, to make his prospecting life a little more com-

----

1 *Canadian Forum*, June, 1926.

fortable, to purchase a tent to sleep in, or an outboard motor for his canoe, or even to build a comfortable cabin to live in during the winter while he tended his trap line.[2] (An alternative way to spend a handy little sum, of course, was to get down to Pine Ridge by the fastest means possible and blow it all in a few days on what seemed, by bush standards, high living.)

Even these small profits, though, made their owners far better off than most of those who rushed in to Red Lake. Most of these men are nameless and faceless to us, but one entered popular history in a dubious fashion. James Swimmings, described as an "itinerant labourer", defended himself on April 1, 1926 before a Montreal magistrate on charges of vagrancy by blaming his condition on the "lure of gold." He told the magistrate that he had just returned, flat broke, from Red Lake. "I didn't strike a paying claim," he expostulated, "I mushed through it all and saw all the hardships, and here is my reward. I am charged with having no visible means of support. I've worked to make a fortune and lost out." The doleful Swimmings was dismissed by the magistrate with "Better luck in future."[3]

The total amount of money expended by the men who rushed to Red Lake depends, of course, on the number who went, and that number is impossible to ascertain. No records were kept of the people who went to Red Lake. (A petition was circulated in March, 1926, among the men at Red Lake that fire patrols be trebled for the coming summer, and that everyone entering and leaving the area sign a register so that a head count could be made at any point.[4] No such record, however, was kept.)

The only reliable record to throw any light on the question of numbers, concerns the number of claims staked during the period. About 6,000 claims were staked in 1926 in the Red Lake

----

2 *Manitoba Free Press*, May 27, 1926

3 *Manitoba Free Press*, April 2, 1926.

4 *Manitoba Free Press*, March 2, 1926.

Mining Division. (This included the Woman Lake area to the east of Red Lake, where gold finds in the late spring and summer of 1926 diverted some of the interest from Red Lake itself.) Every licensed miner was entitled to stake a total of 9 claims, 3 for himself and 3 each for 2 others. But no doubt many claim-stakers did not stake claims for anyone but themselves; and there were many men who went to Red Lake during the peak of the rush and who later staked no claims at all, because all the accessible land was already taken. Considering all of these factors, together with estimates made at various points both by prospectors on the trail and by visiting journalists, a reasonable estimate of the number of people who went to Red Lake during the gold rush is around 5,000 (although the actual number may well have been higher). If 5,000 people rushed in, about $2,500,000 was spent by them. Even if only 3,000 people participated in the gold rush, the amount of money spent on that frantic migration to Red Lake and equally quick exit, was about a million and a half dollars, which in the context of the Canadian economy in 1926, was a substantial sum.

No immediate consequence followed that very large expenditure of money, time and energy. (The amount of money spent on exploration and preliminary development of properties during the gold rush was not included in the above estimates. This amount, which was probably somewhere between $100,000 and $200,000, was not a totally wasted sum, as was the other.) The Red Lake gold rush began in earnest around Christmas of 1925, five months after the initial gold discovery near the shores of Red Lake by Lorne and Ray Howey. Ironically, in those five months before the rush began virtually all of the good ground had been staked, quietly and undramatically, by other members of the large floating group of prospectors who drifted from one camp to another and who were the first to hear of any new finds. And by Christmas of the following year, 1926, it was all over. The water and forest at Red Lake were quiet once again.

Shortly after that, however, new developments occurred in the ownership and direction of the Howey property, and in 1930 the Howey mine went into full operation. Other mines in the area came rapidly into production during the 1930s, several of them major operations which produced gold steadily for 30 years. Two gold mines in the Red Lake area that began production in the 1940s, Campbell Red Lake Mines and Dickenson Mines, are the only two gold mines active in the Red Lake area today.

In addition to the development of the area as a major gold producer, it is possible to outline some other direct results of the Red Lake gold rush. Because the area was so isolated and was such an unprecedented distance from the railway, the rush to get men and equipment to Red Lake was responsible for many innovations in transportation. The innovative use of World War I aircraft to bring men and supplies to Red Lake marked the beginning of the major role which aircraft was to play in the development of Canada's north. One of the quirks which made this gold rush so unique was that it was the first and the last gold rush in which dog teams were the principal means of transportation. (The bizarre combination of airplanes above and dog teams below was one that caught the attention of many contemporary observers.)

Later on, when the Howey mine came into production, an innovative system of marine railways was put into operation for transporting freight over the 180-mile chain of lakes and rivers that connected Red Lake to the railway. This system allowed barge-loads of freight to be taken over the numerous portages on the route, and represented an enormous saving in both time and money. In general, the several innovations in transportation showed that an economically feasible operation could be developed at a greater distance from the railroad than had previously been believed possible.

It was the impetus given by the Red Lake rush to the aerial development of northern mineral areas, in fact, that made the Red Lake gold rush the last to occur in Canada. The last Canadian rush immediately preceding the Red Lake finds (to the Cobalt/Porcupine areas) had been made largely by train. And after Red Lake, aerial prospecting became so efficient and so sophisticated that there was no longer any occasion for large numbers of untrained men to rush off, with much hope but little foresight, to some wild uninhabited district.

Another obvious and important effect of the Red Lake gold rush was the enormous amount of gold which the mines, eventually developed at Red Lake, pumped into the Canadian economy. Up to the end of 1978, the gold mines in the Red Lake district had produced over $630 million in gold bullion (and by far the greater part of this was produced during the period when the price of gold was pegged at $35 per ounce, about a tenth of its current low price). In 1979, 1980 and 1981, Campbell Red Lake alone produced over $300 million in gold bullion. (This was at an average price of $547 per ounce.) These production figures attest to the very large positive contribution which the mines at Red Lake have made to the Canadian economy in terms of actual gold produced, but they do not take into account the many millions of dollars paid in mining taxes by mines in the area, or the very large sums paid in dividends to shareholders.

The statistics of gold-mining production can be used to provide some interesting comparisons. They show, for example, that in 1974, for the first year since statistics were compiled, the gold production of the Red Lake-Patricia area surpassed that of the much more famous Porcupine area (295,000 ounces versus 252,000 ounces). Statistics also support an Ontario Department of Mines publicity piece used in the 1920s and 1930s which proclaimed that Ontario's gold mines were "GREATER THAN KLONDIKE!" The total production of gold from the Klondike fields from 1897 (the first year of substantial production) until the mid-1920s (when production had dwindled to 70,000 ounces) was 8,793,064 ounces of gold. The gold mines of Ontario produced 3,194,308 ounces of gold in one year alone, 1941. (That year represented the historical peak of gold production in both Ontario and Canada.) The gold mines at Red Lake alone produced well over the Klondike total in the 25-year period from 1955 to 1980.

The results of the large contribution made by the Red Lake gold mines into the economy of Canada were more diffuse than those of the Klondike and the Porcupine rushes and their subsequent developments. As a result, it is not possible to point to any sudden spurt in development in another centre as a direct result of developments at Red Lake.

Similarly, the long-term results at Red Lake itself of that dramatic gold rush are not what might be expected from the scale of gold production. In an urban centre, one major industry, on the scale of just one of the big gold mines in Red Lake's history, is responsible for considerable associated economic development, as new smaller businesses, new jobs, new urban developments flow from it. At Red Lake, such associated developments did not occur. In part, this was due to the isolation of the area. In part, this is a problem common to the mining industry as a whole, as mining tends to produce royalties to the government and dividends to the shareholders without significantly enlarging the host community. And all mining centres, unlike other centres that rely on agriculture or manufacturing industries for their economic base, owe their existence to a finite, non-renewable resource and therefore have a life-span measured in relatively few years.

As a result most mining communities have a ramshackle, temporary, haphazard look about them: the town of Red Lake itself, the service centre for the area, is typical in this sense, though there are two "planned" communities in the district. A number of service businesses were developed to serve the mine employees, but, at Red Lake, even this growth was less than it might have been. Five separate townsites developed around Red Lake and in some of these "company towns" the development of service businesses was strictly controlled, or even prohibited, by the all-powerful mining company.

Mining communities follow a growth pattern unlike other communities, for in mining towns a boom-or-bust pattern is always evident. There have been five "booms" in Red Lake's history. The first was the gold rush itself. The second was the "depression boom" when five new mines were brought into production. The third saw three new mines established in the mid-1940s. The fourth, in the late 1950s and early 1960s, saw much exploration and staking in the area but few concrete results. The fifth and most recent, inspired by the unprecedented $800-an-ounce price of gold in 1980, also saw a great deal of staking and exploration, but with the only substantial results being expansions to existing facilities at Campbell and Dickenson, the latter still in the planning stage.

In addition to the Campbell and Dickenson mines, the Griffith Mine owned and operated by Stelco at Bruce Lake, some 40 miles to to the south, has produced about 1 1/2 million tonnes of high-grade iron ore pellets per year, and employed about 500 people when in full operation. Its operation is absolutely dependent, though, on the steel industry in southern Ontario.

Two other major industries are currently active in the Red Lake area, the tourist and the forest industries. Tourism in the summer months began in earnest with the opening of Highway 105 in the late 1940s, and it is now essential to the area. Forest-related jobs currently support more than 300 people in the area. Two major companies are presently harvesting the stands of pine, spruce and balsalm, and huge log-laden trucks are a common feature on Highway 105, rumbling down to the mills at Dryden and Kenora.

The future of the Red Lake area, and its population of some 6,000, is uncertain. Campbell Red Lake, a wealthy gold mine with large proven reserves and a very high-grade ore, should have a stable future for some years if the price of gold does not drop significantly. Although Dickenson Mines has recently encountered some financial difficulties, it is always possible that other mineral deposits may be found which will revitalize the area's economy. Nevertheless, it is highly unlikely that the area's small towns will ever significantly expand into the attractive but agriculturally worthless areas of rock, water and forest that surround them. Red Lake remains a geographically isolated area, even though airplane service and roads link it to larger centres. Red Lake began its historical life as an outpost of the fur-trading companies two centuries ago and, in spite of the gold rush and the development that followed, an outpost it remains.

The Fur Trade Years

Recorded history for most of Canada begins with the fur trade. Until white men penetrated into the wild, vast forested areas of Canada in search of fur, to make money for themselves and for the European nations which were expanding their colonial empires, the New World remained a shadowy unknown. These early, rudimentary commercial contacts had a lasting effect on present-day Canada, for after the era of fur-trade was over, the routes that the traders had established remained. The paths which settlement followed across Canada, before the CPR was built at the end of the last century, were the paths of the fur trade. And even in this century, routes used to penetrate areas like Red Lake have been those used by the fur traders.

The fur trade in Canada depended on water transportation. The surface area of Canada encompasses almost half of the world's fresh water, and the closely connected rivers and lakes of Canada are easily navigable in a portable craft, like the canoe. The surface of Canada is divided into three extensive drainage basins: the St. Lawrence basin, which includes the Great Lakes and the land draining into them; the Hudson Bay basin, including all the land that eventually drains into Hudson Bay; and the Mackenzie River basin, draining into the Arctic. Although Red Lake lies within the Hudson Bay basin, it is close to the rim which divides the Hudson Bay basin from the St. Lawrence basin. This particular geographical location is responsible for the pattern of the pre-gold rush history of Red Lake; and it is also responsible for the importance of transportation problems during the gold rush itself, and the way in which the area was subsequently developed.

The area surrounding Red Lake was first inhabited by humans, now known as Paleo-Indians, around 7500 B.C. From that time until about 500 B.C. two groups of Stone Age people lived in the area, the Shield Archaic and the Plains Archaic peoples. These natives lived on fish, game and edible plants, and probably used copper tools. The following period, from about 500 B.C. until around 1750 A.D., which witnessed the native's first contact with white men, is known as the Woodland Era.

During this period the people gathered together in small encampments and began to use pottery and other ceramics. There is, of course, no written record from this time period.

The beginning of written history for the Red Lake area, as for most of northern Canada, occurred after the formation of the Hudson's Bay Company. In 1670, as a result of two young adventurers' exploration north to Hudson Bay and their subsequent return with a rich cargo of furs, a charter was granted by King Charles II of England for the formation of a new corporation. This company, the Governor and Company of Adventurers of England trading into Hudson Bay, was granted the sole right of "trade and commerce" in all of the land draining into Hudson Bay.

By the time of this charter, the French had already established their hold on the St. Lawrence River and the route which it gave them into the heart of the continent. Independent fur-traders, or "pedlars", both French and Scottish, followed this route. In 1670, with the formation of the Hudson's Bay Company, two competing areas of fur trade were established, following the geographical patterns of the land. One was the vast northern area draining into Hudson Bay, where the Hudson's Bay Company traded for furs, and the other was the wide area to the south, where the independents traded. Red Lake was very close to the boundary of these two areas.

At first there was no conflict. However, the Hudson's Bay Company soon established five forts or "factories" at the mouths of rivers flowing down to the shores of Hudson Bay. A policy of river-mouth trading was established, which required that Indians come from inland to the Company posts with their furs, rather than the Company travelling inland to get them.

For close to a century this proved a satisfactory arrangement. The Hudson's Bay Company was able to turn tidy profits for its shareholders in England by using this stable pattern of operations in Rupert's Land, as the North American territory was then known.

During this period Red Lake remained inhabited only by the Ojibwas, who carried their furs out of the interior to the trading posts. The occasional white trader from the south-east may have visited the area, but these visits are not recorded. A map of the English fur trade in 1700 shows Lake St. Joseph as the approximate northern limit of the trade area of the French, stretching north and west from the Great Lakes. Red Lake, just to the west of Lake St. Joseph, is shown on this map as being in the Fort Albany hinterland (that is, in Hudson's Bay Company territory). Another map of 1750 gives the same information.

However, this happy division of fur-trade areas, between the Hudson's Bay Company and the indepenent "pedlars", was not to last. And when the two forces became direct rivals, marginal areas like Red Lake suddenly came into the picture.

There is no exact date for the formation of the North West Company, but it is normally placed around 1779. This company was, in fact, a loose syndicate of various independent fur-trading firms and individual fur traders, whose trading area extended out from the St. Lawrence basin. At first, the western boundary for the aggressive independent traders was the Lake Superior region. But these traders soon moved west to the valleys of the Red, Assiniboine and Saskatchewan Rivers. And after Peter Pond of the North West Company reached Lake Athabasca in 1778, and Alexander Mackenzie journeyed to the Arctic and Pacific Oceans, an immense northern empire was opened to the North West Company.

This was the "El Dorado" of fur-trading in North America. In the Arctic climate the fur-bearing animals grew thick, lustrous coats, and the territory was virtually untouched, in terms of trapping. The North West Company did spectacular business.

And yet, to bring those rich northwestern furs to market required an incredible transportation system. It was 3,000 miles from Athabasca country to Lachine, Quebec, where ocean-going ships could depart for Europe with cargoes of furs. Unfortunately, there were only 5 months a year when this whole route had open water. So a half-way exchange system was set up. Every year in May, long brigades of loaded canoes set out from either end of the system: those from Fort Chipewyan, laden with furs, and those from Lachine, with trading goods and supplies. The two groups would meet at Grand Portage (in 1803 the turn-around spot was shifted to Fort William). A month was allowed for the turn-around: to repackage goods, to arrange shipments for posts along the way, and to rest the crews. By the end of July, the two sets of voyageurs were on their way home again, one set, to Lachine with furs and the other, to Athabasca with goods.

This transportation system was very expensive, but the great volume of business it supported dwarfed the Hudson's Bay Company's trade by comparison. In 1795, the North West Company controlled 11/14 of the fur trade in Canada, the Hudson's Bay Company 2/14, and independent traders 1/14. In 1800 the Hudson's Bay Company imported into England, furs worth 38,463 pounds sterling. In the same year, the Nor'Westers imported into England, furs worth 144,300 pounds sterling.

At first, the representatives of the Hudson's Bay Company thought that the situation would sort itself out and that they would not have to change their system of river-mouth trading. But by the 1760s, before the actual formation of the North West Company, aggressive independent traders were penetrating northward from the main East-West route (from Lake Superior through Rainy Lake and Lake of the Woods to Lake Winnipeg). These pedlars traded harshly with the Indians but were generous with both credit and liquor (a policy different, in all respects, from the Hudson's Bay Company's), and the representatives of the Company felt that this system of trade was making the Indians lazy and insolent. However, even after the umbrella organization, the North West Company, was formed, the pedlars seemed to be so "free-wheeling" and erratic in their practices, and to have such an unstable organization, that the Hudson's Bay governing committee believed that the new organization would collapse of its own accord.

The traders who worked for the Hudson's Bay Company, however, saw the increasing competition as much more of a threat than did the governing committee, and they finally managed to persuade the Company's administration to take the problem seriously. The Company decided that it was necessary to adopt a policy of expansion to meet the challenge of the Nor'Westers, and decided to do so by pushing westward from Fort Albany. Osnaburgh House was established at Lake St. Joseph in 1786 as the base for exploration of the large territory to the west and south.

In 1787, the North West Company was reorganized and expanded. The Hudson's Bay Company decided to follow its rival's policy and establish a greater number of temporary outposts. To this end Edward Jarvis, the factor at Albany, sent plans to the Committee in the following year, 1788, for a series of new, small posts radiating out from the established Osnaburgh post. These new posts, according to Jarvis' plans, were to be situated at Sturgeon Lake, Red Lake, Portage de l'Isle and Rainy Lake. (These new posts were classified, in fact, not as "posts" but as "outposts": that is, they were considered to be temporary in nature, and therefore, had fewer staff and a smaller volume of trading goods than did the "posts". It was not until the gold rush in 1926, in fact, that the Hudson's Bay Company establishment at Red Lake was given "full post" status.)

Neither the Hudson's Bay Company nor the Nor'Westers considered that these new establishments would be significant in terms of profit. The whole area between the bottom of Hudson Bay and the top of Lake Superior, known variously as "Fire Country" (presumably because of its frequent forest fires) and the "Little North" (in contrast to the "Grand North", Athabasca Country, of the North West Company) generated relatively little trade.

The importance of establishing fur-trading posts in this area was strategic: to prevent the opposition from expanding its territories, and to consolidate fur-trading territory that was near, though not on, major routes. (For the Nor'Westers, the route in question was the main east-west one from Lake Superior, westward through Rainy Lake and Lake of the Woods to Lake Winnipeg. The Hudson's Bay Company was primarily concerned with a proposed alternative to their main path into the continent. Their main route was from York Factory to Norway House, at the north end of Lake Winnipeg, and from there south to Red River country or west to Cumberland House. The proposed alternative route which would approach Red Lake, was from Albany via the Albany River to Lake St. Joseph, Lac Seul, the English River and the Winnipeg River to Lake Winnipeg.)

As a result of Edward Jarvis' recommendations to the governing committee, two new outposts were set up by the Hudson's Bay Company in 1790: at Cat Lake, about 100 miles, by air, north-east of Red Lake, and at Red Lake (sometimes referred to as Red Paint Lake in early journals.) The first documented existence of a Hudson's Bay operation at Red Lake, then, dates from 1790.

The name of Red Lake appeared in written history in two other places at this time. There was a reference to Red Lake, along with the first published story on the origin of the name, in a book published in England in 1791. The unwieldy title of this volume was "Voyages and Travels of an Indian Interpreter and Trader, Describing the Manners and Customs of the North American Indians, with an Account of the Posts situated on the River Saint Laurence, Lake Ontario, etc. . . . ".

The author was a trader named John Long who originally came to Canada from England in 1768 as an articled clerk (presumably with the Hudson's Bay Company). He stayed with the job for seven years, but seems to have been of a more adventurous spirit than is the average clerk. His next occupation was with the army as "a volunteer at the head of a party of Indians" in some skirmishes against the Americans. After that he became an independent fur-trader, and used the whole area north of Lake Nipigon for his wintering-ground.

Long's book was designed to feed the enormous appetite in Britain, at that time, for information about the noble savages who inhabited the continent of North America. In it he gives some fascinating descriptions of contemporary Indian culture, treating us to some details of the harsh practices with which these Indians tested their stoicism. Long concludes, however, that "Notwithstanding the cruelty of Savages, they possess virtues which do honour to human nature, and exhibit instances of generosity and kindness which the most philanthropic soul cannot exceed."[5]

Long was something of a linguist, in addition to his other talents, and his book includes "A Vocabulary of the Chippeway Language", "A List of Words in the Iroquois, Mohegan Shawan-

[5] John Long, *Voyages and Travels of an Indian Interpreter and Trader*, page 119.

ee and Esquimeaux Tongues", and "A Table, Shewing the Analogy Between the Algonquin and Chippeway Languages".

Whether Long actually travelled to Red Lake to trade with the Indians is uncertain. He does give us, though, an account of the origin of the name, which runs as follows.

"A few days after my return to Lac la Mort, a band of Savages arrived from the Red Lake, called by the Indians, Misqui Sakiegan . . . Red Lake is so called on account of a remarkable circumstance which happened to two famous warriors of the Chippeway nation, who were hunting by the lake side, and as they were looking out for game, perceived at some distance an enormous beast, that appeared much larger than any animal they had ever seen; his pace was slow and heavy, and he kept constantly by the water side. They followed him as close as they thought prudent, determined at all hazards to use their best endeavours to kill him. As they approached, they had a clearer view, and discovered that his body was covered with something like moss; this increased their surprise, and after consulting together, they continued advancing towards the beast, and fired large shot, without appearing to make any impression. They fired again with as little effect as before; then retreated some distance, sat down and sung their war songs, addressing themselves to the Master of Life, and desiring his assistance to enable them to conquer it, as they believed it to be the Matchee Mannitoo, or bad spirit, in the shape of this monster. They then got up and pursued him, both firing at the same time: the shot proved successful, and caused the animal to turn around, which induced them to keep up their fire till the beast jumped into the water, and they lost sight of him. From the circumstance of his blood dying the water red, this lake has ever since been called the Red Lake."[6]

Long also gives a description of the country around Red Lake, although it is again uncertain whether this is the result of his own observations or of reports from his Indian trading customers. He says of the Red Lake area that "Fish is caught here

----

6 John Long, *Voyages and Travels of an Indian Interpreter*, page 80

in abundance, and wild rice grows in very great plenty in the swamps. The country likewise abounds with all sorts of animals for hunting . . . the Indians are very fond of fishing and hunting here during the winter season, as they are generally very successful even in the most severe weather."[7]

At the same time as that described by Long, that is, in the years just before the Hudson's Bay Company established their outpost at Red Lake in 1790, the North West Company established a trading post at Red Lake. The North West Company post there was usually referred to as Red Lake House and it was established, it seems, in 1786.

The first post of the Nor'Westers seems to have been located on the shores of Red Lake near the mouth of the Chukuni River (the southern link of the lake to the English River system and beyond). At Red Lake House, as with all other small posts in the area, operated by either company or independent trader, it was the custom for the post to be inhabited by the trader only for the trapping season each year, from fall to spring. The Hudson's Bay Company Archives contain a reference to the North West Company's resident trader as a "Mr. Cameron", but it is not known whether this was the Duncan Cameron who described the area's Nor'Westers' operations in his journals.

This Duncan Cameron entered the employ of the Company as a young man, becoming a clerk in 1786 and being placed, then, in charge of the "Nepigon" region, which covered the vast tract of land north and west of Lake Nipigon and included Red Lake. The district was not a wealthy one in terms of furs, and it seems that Cameron worked hard to make it pay.

Cameron, like John Long, turned his wilderness experiences into print for the benefit of the eager audiences at home in Britain. He published two books: *Nipigon Journal* and *Sketches of the Natives of the Nipigon Country*. As the clerk in charge of the Nipigon region, he made a number of journeys into the interior, and in his journal of 1802 Cameron describes one such visit which took him to the Hudson's Bay Company's Osnaburgh

----

[7] John Long, *Voyages and Travels of an Indian Interpreter*, page 81.

post. He reports a conversation during his visit as follows:

"This man (an Indian from Red Lake) tells me that the English (the H. B. C.), the XY, and Mr. Adhemar (a free trader) were striving who would squander the most and thereby please the Indians best."[8]

This entry in Cameron's journal provides, in fact, one of the few existing pieces of evidence to show that the XY Company also traded at Red Lake. This fur-trading firm was a short-lived one. The XY Company was formed by the union of two existing companies, in 1798, in opposition to the existing North West Company. The official name of the new firm was the New North West Company. But, because its initials were so similar to those of its rival firm, the new company's goods were labelled with the initials "XY" rather than "NNWC", to avoid confusion in transit. The name XY survived, but the firm did not. In 1804 it was absorbed into the (old) North West Company.

Although the XY post at Red Lake was short-lived, the two larger rival companies kept pace with each other for a period of some thirty years, until the Hudson's Bay Company and the North West Company merged in 1821.

The obvious effect of this competition for furs on the trading companies was to make operations, in marginal areas like Red Lake, even more financially precarious for each company than they would otherwise have been. The effect of this competition on the Indian population, however, is not so clear-cut.

The period of intense rivalry for furs in the area around Red Lake, lasting from around 1785 to 1821, was also a period when big game was plentiful in the area. As a result of these two factors it was a relatively prosperous time for the Ojibwa population. The role that gifts of liquor played in the fur-trade, at that time, was a major one. We know that independent traders, North West Company traders and Hudson's Bay Company traders all traded alcohol for furs, and gave alcohol as a bonus as well. But different sources give different accounts of the

----

8 Quoted in George Bryce, *The Remarkable History of the Hudson's Bay Company*, page 183.

importance of this aspect.

For example, Nor'Wester Duncan Cameron describes, in scandalized detail, the conduct of the Hudson's Bay Company men at Osnaburgh in this respect. He says that "there are 8 Indians here, all drunk and very troublesome to my neighbour [the HBC trader] who, I believe, is as drunk as themselves", but goes on to admit that, "I am giving them plenty to drink without getting anything from them yet." And he reports a situation at Red Lake, where a four-way competition for furs was such that, "an Indian who has no influence got 5 kegs of mixt wine to himself and credit for 200 skins . . . all the Indians were 20 days without getting sober."[9]

Other sources, however, defend the Hudson's Bay traders with regard to the destructive trade in alcohol, saying that the Nor'Westers were traditionally much more generous with liquor than the competition. The two major companies' basic policies of payment were different: the Nor'Westers were flexible about prices for furs, though they generally paid on the low side, and were easy with credit and with gifts; the Hudson's Bay Company had a fixed price for furs in each district, somewhat higher than the Nor'Westers' price, and gave "presents" only to those Indians whose favours they particularly wanted.

It is doubtful whether the results of these two policies were much different, as far as the Indians were concerned. The Indians moved around freely from post to post to take advantage of the best prices. They were described as being hard bargainers, "acutely aware of variations in price and quality", and apparently they would "undertake long journeys to secure better bargains. When they thought they were being unjustly treated they would refuse to trade at all."[10] In spite of these benefits, many of the Indians fell into a credit trap whereby they would spend the whole season's proceeds before they earned it, year after year, and, in effect, would be working for nothing.

----

9 Quoted in George Bryce, *The Remarkable History of the Hudson's Bay Company*, page 18.

10 E.E. Rich, *The Fur Trade and the Northwest to 1857*, page 92.

In the long run, it seems that the fur trade did no favours for the Ojibwa who inhabited the area that included Red Lake. When the competition for furs ended in 1821, the Indians found they had only one company to deal with; that one company of course could dictate the terms on which furs were traded, and it was located at some distance from them. Their area was depleted not only of furs but of most big game by 1821, and the Indians' introduction to liquor had far-reaching and tragic consequences.

But the trader-Indian bargain looked quite different from the opposite point of view. The Indian had not only shown the trader where to go to get furs (most of the fur-trade routes had been used by the Indians for centuries). He also showed the white man how to get there: the canoe, the snowshoe, the toboggan and the dog team were all Indian methods of transportation (and the very methods of transportation, as well, that brought most of the people to Red Lake during the gold rush).

The period of competition for the Indians' furs at Red Lake was also the period that was documented, in most detail, by the companies in question. It is possible to compile, from journals and other records of both the North West Company and the Hudson's Bay Company, a fairly detailed account of activity at their posts on Red Lake during these years.

On August 13, 1790, a Hudson's Bay Company trader named James Sutherland arrived at Red Lake from Osnaburgh House to establish an outpost. Sutherland was a tailor who had been in the service of the Company since 1770. He was apparently a soft-spoken, thoughtful man of deep religious beliefs who baptized children and frequently held Divine Service on Sundays.[11] (In all these respects, he presented a sharp contrast to the majority of his fellow traders.)

Three days after his arrival, he began laying the foundations for his house, which seems to have been near McNeely Bay, a small bay on Red Lake, about a mile north of the Chukuni River outlet. He also filed a plan of it; the plan shows his building to have measured 38 x 18 feet, of log construction with

----

[11] Rae Kiebuzinski, *Yesterday The River,* page 61.

six very narrow windows and two fireplaces. It was divided into three sections: a private room for his own bunk and office, living quarters for his men, and a storeroom.

The rival trader for the Nor'Westers, who had been trading at his post, Red Lake House, since 1786, was not yet in residence for the season when Sutherland arrived. When he did return, he built a new house for himself close to Sutherland's, and cordial relations were established between the rival operations. The competition for furs was keen but the traders remained friendly: Sutherland recorded that they played football on the lake ice on New Year's Day, 1791, and that, on May Day, the Nor'Wester set up a maypole in front of Sutherland's house.

In the fall of 1791, John Best replaced Sutherland at the Hudson's Bay Company's Red Lake outpost. He had the same friendly relationship as his predecessor with the Nor'Westers, and spent the trading seasons at the post until the spring of 1794. At this point Magnus Birsay took over the Hudson's Bay Company post for the summer. His Nor'Wester rival at that point was one Venant St. Germain. In the autumn of 1795, though, the Red Lake outpost was abandoned in favour of the one at Escabitchewan (on Ball Lake, some 60 miles by air south of Red Lake), presumably because the intake in furs had not been great enough.

In 1799 Magnus Birsay returned to Red Lake to re-establish the Hudson's Bay Company outpost. He found the Nor'Westers occupying his old house, so he built a new one. Motivated, perhaps, by a desire to keep up, the Nor'Westers did the same. The Hudson's Bay post was abandoned for a year at the end of that season, rebuilt on another site by Thomas Vincent in the summer of 1802, and on yet another site by Vincent in 1803.

In 1804 William Thomas came in to replace Vincent at the post as head. He found the house burned down when he arrived, and so built another a mile away. In that year the Nor'Westers, the Hudson's Bay Company and the XY Company all traded at Red Lake. (This was the year in which the XY Company was absorbed by the North West Company. But apparently news of the union didn't reach Red Lake until February of 1805.)

1804 was the year in which a most dramatic episode is said to have occurred at Red Lake. (There is, unfortunately, little evidence anywhere to support this story; it comes from John McKenzie, apparently, who worked for the Hudson's Bay Company, in various capacities, from the early years of this century to the 1920s in northwestern Ontario.)

It is known that the area directly south of Red Lake, where the standard east-west transportation route for both fur-traders and Indians was found, was an area hotly contested by a number of Indian tribes in the 18th century. The Indians received firearms from all the trading companies which they employed in their inter-tribal warfare. The story in question tells that one Donald McKenzie, who had a trading post on the largest island in Red Lake, was warned by a runner one day in 1804 that a large war party of Sioux Indians from the west was coming in his direction. He and his small party made haste to flee, but before doing so, dumped some 80 ten-gallon kegs of rum into a small lake on the island. The story does not have a happy ending, for the McKenzie party was caught and massacred near Sioux Lookout (so named because it had a height of land which was used as a lookout for approaching Sioux war-parties). Nor has a positive twist been added to the story in contemporary times, for two local skin divers searched both small lakes on McKenzie Island during the 1960s without finding any rum.[12]

The 1805-1806 trading season marked the climax of fur-trading in the Red Lake area. In that year William Thomas was in charge of the Hudson's Bay Company, with a surprisingly large staff of eleven men. John Haldane headed the North West post there, with a force recorded at an astounding 54 men in total. Whatever the reason for the large groups of traders, there was no safety in numbers. The friendly spirit had gone. In the spring of 1806, the Nor'Westers looted the Hudson's Bay Company house, and, apparently, many other unpleasant incidents occurred as well. As a result, no-one wanted to work there any longer, and the Hudson's Bay Company post was closed in the spring.

----

[12] D. F. Parrott, *The Red Lake Gold Rush,* page 14.

The only evidence of any trading activity in the Red Lake area over the next ten years is a North West Company journal entry to the effect that, in the fall of 1807, Haldane again went to Red Lake.

In the fall of 1815, the Red Lake area once again became of interest to the fur-trading companies. At that time James Slater, who was in charge of the Escabitchewan post, decided that since the immediate area surrounding Escabitchewan was exhausted of furs, and since he had heard that the North West Company was sending men to spend the winter at Red Lake, he would send someone there as well. Accordingly, he despatched one of his employees, Edward Mowat, to winter there.

In the following season, 1816-1817, the Red Lake District was formed by the Hudson's Bay Company. It was composed of the posts at Red Lake and Escabitchewan, and was under the direction of James Slater. Slater went to Red Lake himself for the season, and built new buildings (again) for the outpost. Again, there was unpleasant rivalry in the 1817-1818 season between the Hudson's Bay Company traders and the Nor'Westers, whose buildings were only a few yards apart. Mowat wintered at the Hudson's Bay post in 1818-1819, followed by Marcus Calder in 1819-1820. The following season, Slater had insufficient provisions to outfit an outpost at Red Lake, and the post did not open.

The last report of any fur-trading during this period at Red Lake is from the Hudson's Bay Company files. A report on the Osnaburgh District in 1822 records the arrival of Mowat, Calder and nine men from Red Lake and Escabitchewan, with 32 packs of furs and 130 gallons of rice, indicating that the Hudson's Bay Company post at Red Lake was occupied during the 1821-1822 season.

It was in 1821 that the rivalry between the two major companies ended. The rivalry had come to a head when Lord Selkirk, who had a controlling interest in the Hudson's Bay Company, established the Red River colony which cut across the Nor'Westers' lines of communication to the west. The Hudson's

Bay Company traders had invaded the rich Athabasca country as well. The Nor'Westers, whose enormous wealth had gone more to the partners than to the company, had no money to pay for the expensive and prolonged lawsuits that followed. Although they won in the courts, they lost over all. In 1821, the North West Company was absorbed by the Hudson's Bay Company.

After the rivalry was over, there was no need to maintain the temporary outposts which had been established mainly for strategic purposes (that is, because the opposition had an outpost there which the local Indians would naturally use rather than journeying elsewhere with their furs). The fur trade in the Red Lake area had never been described in glowing terms. A typical statement about the area, from the Hudson's Bay Company files of 1816-1817, refers to the "now exhausted state of the Country, particularly in Beaver."[13] When news came of the amalgamation of the two rival companies, then the Hudson's Bay Company's Red Lake outpost was abandoned, at the end of the 1821-22 season.

This marked the end of an era in Canadian history, and the end of a chapter in the history of Red Lake. Red Lake was no longer important to the fur trade, and the fur trade itself was no longer to be important in the exploration and development of the vast unknown parts of Canada. Other forms of wealth would spur the next stages of frontier development: agricultural wealth, in the rich soil of the Prairies; and mineral wealth, in the Cariboo, in the Klondike, in Cobalt/Porcupine, and in Red Lake.

After Mowat and Calder brought their 32 packs of furs out from Red Lake and Escabitchewan in 1822, there is no solid evidence of any fur-trading post at Red Lake for almost a century. The Red Lake outpost appears on several Hudson's Bay Company maps during this period, but there is not a single reference in Company archives (or elsewhere) to a trading post at Red Lake until 1918. In that year, it reappears in Company documents as an outpost of the Pine Ridge post on Lac Seul. The location of

----

[13] Hudson's Bay Company Archives, B.177 / e / 1, fo. 3d.

the outpost, at this time, was at Post Narrows, near the northeast end of the lake, some 8 or 9 miles by water from the old location further south at McNeely Bay. The reason for reopening the outpost in that year was, again, a strategic one: George Swain, a free trader who appears later in this story, had established himself in the area near Gullrock Lake, and Indians in the area were bringing their furs to him rather than taking them all the way to the Hudson's Bay Company's Lac Seul post.

However, during the blank years between 1822 and 1918, there are references to a Lac Seul trader, John Setter, trading with Indians from Red Lake in the 1880s. There is also another reported piece of evidence which fits in with few of the established facts, for it suggests, not only that there was a Hudson's Bay Company post near Post Narrows in the middle of the 19th century, but that this post was undergoing severe hostile rivalry with some free traders. Several years ago, two revolver barrels were found buried at Post Narrows. They were apparently identified as belonging to firearms manufactured in 1840 and issued to the British army, and were conjectured to have arrived at that spot in the possession of British soldiers sent to defend a Hudson's Bay Company post from the hostile advances of free traders.

After the outpost was re-established in 1918 (however long had been the interval since it was last in operation), it was a quiet backwater posting for several years. But with the sudden influx of prospectors that followed the Howey brothers' gold discovery in 1925, the outpost's role changed dramatically. Many of the Indians who had been trapping furs to bring to the outpost abandoned their traplines in favour of acting as guides and bearers for the prospecting parties. The Red Lake outpost became a supply depot for the gold-seekers. It was given full-post status and moved to a point nearer the Howey operation.

There has been a Hudson's Bay Company store at Red Lake since then. There are, in fact, two separate "Bay" buildings in the town of Red Lake today: one a supermarket and the other a three-storey department store, the largest in the area. There was a Hudson's Bay store at Madsen from 1950 to 1972, and there has been one at Balmertown from 1953 to the present.

Both the Red Lake and Balmertown operations still buy small quantities of furs from local trappers.

The Quiet Years

When the fur-trading chapter of Red Lake's history ended, it might be said that the area went back to sleep. And while it was asleep, ownership of the land changed hands several times.

The first time this area of forest, water and rock had been claimed by anyone in written history was in 1670. Rupert's Land was the vast section of North America which was granted by King Charles II to the Hudson's Bay Company in that year. It was composed of all the territory draining into Hudson Bay, but its boundaries were never precisely defined. There was no doubt, however, that it included areas as close to the waters of the Bay (and of as dubious economic value) as Red Lake. In the Treaty of Utrecht of 1713, Red Lake is acknowledged to be part of the Hudson Bay region.

And Red Lake remained part of Rupert's Land for 200 years, until the enormous territory was bought by the government of the newly-created Dominion of Canada, in 1869. Canada received Rupert's Land as its own, and the Hudson's Bay Company in return received property rights to the land surrounding its posts, land grants in the Prairies, and a cash settlement of 300,000 pounds sterling.

There was another pressing question at this time regarding land ownership in what is now northwestern Ontario. Construction of the C.P.R. began shortly after Confederation, and it was necessary to establish rights-of-way (and associated timber rights) for the railway company. As a result, two treaties were drawn up and signed with the Indian populations in the area: Treaty 5 (the Winnipeg treaty) for the Berens watershed, draining westwards into Manitoba, and a second treaty, the one concerning the Red Lake area, Treaty 3.

This agreement, called the Northwest Angle Treaty, was signed on October 3, 1873, between the Government of Canada and the Saulteaux (Ojibwa) and Chippewa Indians. It applied to the lands drained by the English, Winnipeg and Rainy Rivers. In return for surrendering their general claims to the land, the Indians were granted gifts and annuities, as well as an unusually large amount of land (probably because the land in question was thought to be of little value).

The two hundred years of activity of the Hudson's Bay Company in this part of Rupert's Land had led to very little mapping of the area. Although Company maps of the early 1870s suggest that a line drawn through the English River and Lac Seul would generally define the northern limits of any mapping, Red Lake, to the north of this, does appear on maps at this time.

The Geological Survey of Canada, a scientific branch of the federal government which had been in existence since 1842, became responsible for exploration and mapping of the extensive new territories added to Canada at this time. Alfred Selwyn was appointed Director of the branch in 1869. One of the men Selwyn recruited for help in the enormous mapping job was Robert Bell, a professor of geology at Queen's who had already been involved in two geological explorations in northwestern Ontario (in the Lake Nipigon area).

In 1872 Selwyn and Bell decided to do a general survey, by canoe, of the region westward from Port Arthur to the Prairies. On this survey they passed through Lac Seul and the English River. It was, apparently, at the Hudson's Bay post at Lac Seul that a group of Indians showed the Selwyn-Bell party some samples of soapstone or "pipestone" (so called because Indians carved bowls for their pipes from the soft stone) which came from Red Lake. This was the first clue to the mineral potential of Red Lake, as soapstone suggests the presence of sedimentary and volcanic (or igneous) rocks, where gold veins are usually found.

On this occasion, as on many others, it seems that Selwyn's Indian guides, and his encounters with other Indians, were invaluable sources of information. This was in spite of a reported Indian belief that an Indian who showed a valuable or unusual piece of rock to a white man would die within a year.

The first flickerings of interest in the mineral potential of the Red Lake area, then, were aroused by Selwyn and Bell's trip of 1872. It was not until ten years later, however, that Red Lake itself was visited by representatives of the Geological Survey of Canada. In the meantime, Bell made further surveys in northwestern Ontario of the Lake of the Woods and Rainy River districts, and of the Albany River to James Bay.

During the 1880s the Geological Survey of Canada directed its attention to the area drained by the English and Berens Rivers. (The Berens River system runs in an east-west direction some 60 air miles north of Red Lake.) In the summer of 1882, Bell made a quick survey of the Red Lake - Chukuni River district, and his report on the survey stated that "its (Red Lake's) shores proved to be of great geological interest. The whole lake (which is of considerable size) lies within a wide belt of Huronian rocks, among which several of the rare varieties are well developed, and they were found to contain some interesting minerals."[14]

The detailed information and optimistic conclusions of Bell's reports fuelled the imaginations of the politicians and businessmen in southern Ontario. At this time, the part of what was formerly known as Rupert's Land that included Red Lake, belonged to the federal government, and many years of federal-provincial and inter-provincial (Manitoba vs. Ontario) wrangling followed, to determine the ownership of this territory that was now seen to have great potential. In 1888, the part of this territory that lay north of the Lake Superior-Huron watershed and south of the Albany and English Rivers was granted to Ontario. More disputes followed and it was not until 1912 that the territory north of the English-Albany Rivers and to the coast of Hudson-James Bay was granted to Ontario, thus completing its present-day boundaries. This new territory, which included Red Lake, was named the Patricia District.

The next geological report on Red Lake was made by D. B. Dowling, also of the Geological Survey of Canada. In 1893 he was sent to make a survey of the route north of the English River through Red Lake, and of the Vermilion River to the Berens River and west into Manitoba. The geological map which Dowling made of this area was the only such source of information for the next 30 years. Dowling felt that the area had a high potential for minerals, though he didn't find any specific deposits. It seems that Dowling was subject to the same impulse that

----

[14] Robert Bell, *Geological Survey of Canada Report of Progress, 1882-83-84*, pt. A, p. 5.

frequently goads contemporary travellers when confronted with the rock faces of Precambrian Shield country, for he noted in his journal that he carved his initials and the date in the soapstone cliff at Pipestone Bay, at the northwest end of Red Lake.

This survey like all the earlier explorations in the area, failed to lead to any significant mineral discoveries in the north. In 1887-88, Ontario had sponsored a commission to study Ontario's mineral resources. As has frequently been the case with government-sponsored commissions, the commission did not bring any new information to light, but it did result in the formation of a new arm of government bureaucracy: in this case, the Ontario Department of Mines, created in 1891.

At this time there was still some belief in, "Old Ontario" (which is today usually termed southern Ontario), that the newly-acquired northern regions would be capable of sustaining large populations. Surveys made in connection with the building of the C.P.R. noted any land that might be used for agriculture. Free land grants to settlers were still operative. In spite of this, the general isolation of what is now termed northwestern Ontario, and the paucity of fertile lands, did not prove very attractive to settlers.

By 1910 it was sadly recognized that there were only three pockets of worthwhile arable land in the whole northwestern area: the first between Rainy Lake and Lake of the Woods, the second near Kenora, and the third around Dryden. One surveyor pessimistically judged that there were only two pockets of decent land in 350 miles, after going west from Fort William. "North and south [of the railway line] the country is a wilderness of lakes, or rather tarns, on a large scale, filling holes scooped out of primitive rock."[15]

In 1911, the total population of Thunder Bay, Rainy River and Kenora Districts, three very large areas, was only 66,000 persons. And this was at a time when Toronto was already a bustling metropolis of close to 400,000.

----

[15] Quoted from G.M.Grant's *Ocean to Ocean* by G.P. deT. Glazebrook, *Life in Ontario: a Social History*, p. 149.

These meagre populations included all those "hopefuls" who had flooded in during the first wave of mineral discovery in northwestern Ontario, a wave whose furthest edge only briefly touched Red Lake. The first gold discovery in Ontario had been made on a farm in Madoc township, north of Belleville, in 1866, and the first significant gold discovery in northwestern Ontario was made near Jackfish Lake, in the Lake of the Woods area, in the winter of 1870-1871.

The northwestern discoveries were stimulated by the sudden jump from Stone Age to Steam Age transportation that occurred when the railway was pushed through, from the Lakehead west to the Prairies, in 1885. No vast ore bodies were uncovered in the course of construction, as happened later in northeastern Ontario, but the railway did make the country more accessible to prospectors and speculators (if not more attractive to the settlers who the government fondly hoped would go north).

The Jackfish Lake gold discovery was the first of many in the Lake of the Woods district in the 1880s and 1890s. At the height of operations in the area, there were a dozen producing gold mines. They were most notable, however, for a losing combination of small ore bodies and inefficient operations. (This combination led to the caustic observation on gold-mining in Ontario in the early years of this century, that it earned a dollar for every hundred it spent.) The rush of speculators to the Lake of the Woods area had brought a total of 50 stamp mills into operation. (Stamp mills were simply mills that crushed ore with iron-shod timbers that "stamped" the ore brought up from underground.)

A report in 1904 on gold-mining in the area, however, said that, of the 50 stamp mills in northwestern Ontario, only 7 ran at all and only three were in more-or-less continuous operation. It seems that many operations in the area were totally fraudulent, with mills constructed where there was no ore and no possibility of finding any. Bad management prevailed, there was inexperienced labour at all levels, and milling techniques were poorly understood. As a result, the report said, 25% to 50% of the gold was often lost in the milling process.

These Lake of the Woods experiences did nothing to improve Ontario's reputation, at the time, as a non-producer of gold. In 1899, the year when the area's production of gold was at its height, Ontario produced a mere 20,394 ounces of gold, about 1/80 of the total gold production in Canada. (In comparison, in 1981 Campbell Red Lake alone produced 200,528 ounces of gold.)

The late nineteenth century's small influx of prospectors and speculative capital into the Patricia District touched Red Lake at only one spot, Slate Bay, about half-way along the lake on its north side. A British company called the North Western Ontario Development Company, represented by a Canadian, R. J. Gilbert, came to Red Lake in the summer of 1897. (It is possible, because of the timing involved, that this company was inspired by Dowling's optimistic report, produced after his visit there in 1893.)

In any case, the early summer of 1897 saw the Gilbert party disembark from Dinorwic, on the C.P.R. line. They paddled their canoes and portaged the distance of some 200 miles to Red Lake. The party made camp and started prospecting in the Slate Bay area, and, apparently, were fortunate enough to find veins containing free gold which could be flaked off and panned.

Exhilerated, no doubt, by the find, the party staked 8 claims in the vicinity and decided to make a quick trip to Hamilton, where Gilbert knew a surveyor, James Tyrrell. He planned to enlist Tyrrell's services to survey the newly-staked claims. As the party was preparing to embark from their camp at Slate Bay, Gilbert reached down to pick up his revolver which lay on the smooth rock near the shore. As he picked it up, the revolver slipped out of the holster, struck the rock and discharged the bullet, striking him in the chest. He died instantly.

Instead of immediately proceeding to Hamilton, the disheartened party instead went only to Rat Portage (Kenora), carrying their heavy burden of Gilbert's body in the mid-summer heat. At Rat Portage, the group purchased mining equipment and supplies and returned, again via Dinorwic, to Red Lake in late September.

Although the party did not go to Hamilton at that time, they obviously did contact Tyrrell, the Hamilton surveyor, as he came to Red Lake before freeze-up and surveyed the claims. The party built two cabins on their claims at Slate Bay and sank a 25′ shaft. This is all that is known, however, about the venture. It is obvious that it was short-lived and equally obvious that it was unsuccessful, probably due to a combination of the extreme isolation of the spot and the fact that the gold-bearing vein at the shaft was of irregular size.

For the next 25 years, there was little activity at Red Lake. This was due to a variety of reasons. First, the stupendous mineral discoveries in the Cobalt/Porcupine areas were made during this time, and therefore, this area drew to it both prospectors and capital in very large quantities. Second, the years of World War I and those immediately following it were not a time when there was, generally speaking, either time or money for exploration, surveying or prospecting in marginal areas like Red Lake. And finally, the isolation of the Red Lake area, and its distance from the railway line, still seemed an insurmountable barrier. It was known that any gold discovered in that area would be lode gold, gold in veins buried deep in rock, gold that would demand expensive, heavy, bulky machinery to first remove it from the ground and then to separate it from the surrounding rock.

As a result, the years between the ill-fated Slate Bay development of 1897 and the small silver finds at Red Lake in 1922 saw only a very small amount of activity in the area. During this quiet time at Red Lake, there were some independent prospectors who roamed the area; there were a few trappers who made their living in the bush, their number augmented after World War I by ex-servicemen who had nowhere else to go and nothing else to do; and there was a fur farm started in 1912 by five enterprising trappers, which remained in business until 1916, when the Hudson's Bay posts in the area stopped buying raw furs.

One occasional visit to the Red Lake area during these years, by a prospector named McDonald, had some interesting results, however. He seems to have spent some time prospecting on the old claims of the North Western Ontario Development Company. When he moved on to the next promising spot (wherever that might have been) he left behind, in his cabin, some papers which showed that he had once been at Cobalt, along with a sample of silver ore with "Cobalt bloom" on it, and a distinct pinkish coating. In 1919 the fur-buyer and trader, George Swain, came upon the cabin, and took away the sample.

Swain was one of the colourful figures who enliven the history of the north. He had an unusual background for a fur-trader, as he had been a medical student in Winnipeg when the combination of his fiancee's sudden death and the onset of pleurisy forced him to stop studying in 1904 and to head for a new life in the fresh air. In 1911 he built the first of his many trading posts in the English River area. He travelled widely in the region, buying furs and wild rice from the Indians, trying his hand at fur farming, and moving enormous quantities of trading goods (often singlehandedly) from one trading post location to another. He was a dramatic figure to encounter, wearing a western-style hat, a cartridge belt, colt revolver, and fringed buckskin pants and gauntlets which he had made himself.

After the Cobalt sample had been sitting in Swain's cabin for a number of years, some Indians saw it and told Swain that there were rocks similar to the sample to be found in East Bay of Red Lake. Swain went to East Bay, looked at the showings, and decided to take the original sample with him to Winnipeg. The sample eventually reached the hands of two experienced prospectors who thought the East Bay formations, which the sample was said to resemble, might well contain silver. They raised $500, went to Red Lake, staked six claims, and took samples back to Winnipeg for assaying. The samples were shown to contain, not just silver, but seven of the nine contained small amounts of gold as well.

The modest silver discoveries at East Bay were immediately hailed by the press as a rich silver find and a number of prospectors, probably no more than 50, paddled in to Red Lake during the summer of 1922. *The Manitoba Free Press* reported, on September 25, that 68 claims in the East Bay area had been recorded at the Mining Recorder's office in Kenora. The fact that assay samples from claims made in the area showed gold as well as silver seems to have made little impact. Gus McManus, then mayor of Hearst, arrived in Red Lake in late September to prospect. Both sides of East Bay were already staked, so, instead, he and his party went to the south-east end of Red Lake to prospect. In a small bay there they found a vein with visible gold, and staked 16 claims. One of the few tangible results of McManus' discovery was that news of it brought Ken McDougall, who was later to become one of the pioneers in transportation to the area, to Red Lake for the first time.

Claims with gold showings were also staked that summer on properties which later produced the McMarmac and Cochenour-Willans mines. Overall, about 100 claims were staked, but with few results. In fact, all the claims staked that summer were allowed to lapse. None of the claim-holders were able to interest any mining capital (and in Canada, this meant any of the Toronto mining interests) in their finds. The grade of ore was too low and the distance of 180 miles from the railroad was too great, as you could go by power boat only about half-way from the railway to Red Lake.

The small revival of prospecting interest that touched Red Lake in 1922 was part of a general improvement in the Canadian economy after the slump that followed World War I. The general upswing also brought a revival of government interest in the area. In 1922 the Ontario Department of Mines decided to continue the geological survey of the Patricia District, and despatched Dr. E. L. Bruce to do a survey of the upper parts of the English River valley. The Red Lake area, in particular, had come to the attention of the Department of Mines because of the reports of mineral discoveries and consequent stakings at the time. As a result, if time permitted, Bruce was to go on to survey Red Lake as well.

Bruce had finished the initial part of his survey by the end of July, so he continued on to Red Lake. He spent about 6 weeks there with his party, and apparently encountered a number of prospecting parties. In any case, he found the area sufficiently interesting that he returned in the summer of the following year, 1923, to do a more complete and detailed survey. Summarizing what he had done in his report, Dr. Bruce said that, "the aerial mapping and geological investigation" of the Red Lake basin was completed. "A micrometer survey was made of Red Lake and of smaller lakes draining into it, so far as they could be discovered and reached, and the country inland was explored by compass and pace traverses, so that the area adjacent to the lake has been examined as thoroughly as circumstances at present warrant."[16]

The report which resulted from Bruce's survey of Red Lake was published by the Ontario Department of Mines, with an accompanying map, in 1924. It is a significant link in the chain of events that led to the gold rush of 1925-26 and the subsequent development of the area as a major gold producer. For it was this report which led Lorne and Ray Howey to look for financial backers and to go to Red Lake to prospect in the following year.

Bruce's description of the geological possibilities at Red Lake could be described as lukewarm: it was the most positive official assessment of the area's potential that had yet been made, but a far cry from the glowing reports that later appeared in the world-wide press. Bruce was at least as much impressed by the beds of clay soil at the east end of Red Lake, at that time cultivated, in part, in the "splendid gardens" at the Hudson's Bay post, as he was by the mineral possibilities of the Red Lake basin. In the section of his report entitled "Natural Resources" (which includes soil, timber, fish and game, water power, and minerals) he remarks, "Of the present and prospective resources of the Red Lake basin, the soil which covers a large part of it so completely is probably the most important."

When he turns directly to the question of Economic Geology and reviews the present situation, he is hardly more positive about mineral development. He notes that "the amount of mining development work done in the vicinity of Red Lake is comparatively small." He remarks on the presence of an old "shaft 25′ deep sunk one-quarter of a mile inland from the north shore of Slate Bay" (this dated from the Gilbert party's short-lived

---

16 E.L. Bruce, *Geology of the Basin of Red Lake, District of Patricia*, p. 12.

efforts in 1897) and notes that some claims were staked after silver was found in the East Bay area in 1922. Of these claims, he says, "most . . . were staked in the hope that interest in the district would make possible the easy disposal of them before it was necessary for any work to be done. As there is no demand for claims, most of the owners have preferred to allow their stakings to lapse. . . . Hence no prospecting whatever has been done over much of the promising area, and thorough prospecting in but a very small part of it." Overall development of the Red Lake basin when he left it, at the end of the summer of 1923, in fact, consisted of 7 isolated cabins; the Hudson's Bay post; and 5 areas where claims had at one time or another been staked.

But when Dr. Bruce, in his report, turns to the geological potential of the area, optimism surfaces. He says that "the geological succession . . . is a promising one for gold-bearing quartz veins, if this area may be judged from other pre-Cambrian areas. With this in mind, a network of quartz stringers [a vein] . . . was sampled. . . . In spite of the favourable appearance of this vein it contained neither gold nor silver on assay. Quartz veins found in claims . . . do, however, contain gold in visible quantities." Dr. Bruce lists the value of 8 samples taken: the three that contained gold did so in the proportions of $6.20 per ton, $19.00 per ton, and $25.80 per ton. He summarizes his findings by saying, "The possibility of the occurrence in this district of minerals in commercial quantities is still a debatable question."

At the same time as the minimal development and the far-reaching geological reporting of the Red Lake area were taking place, there were enormous and rapid developments in gold and silver mining in northeastern Ontario. These were connected closely to the later development of Red Lake. Transportation, of course, was one of the basic requirements for growth. The lack of rapid, modern, easy transportation explains why the area was relatively late in developing; it accounts for the peculiarities of the gold rush; and it led to important innovations in transportation. And transportation was also fundamental, but in a quite different sense, to the discoveries at Cobalt, Porcupine, and Kirkland Lake.

In 1900, Canada seemed to be poised on the brink of becoming one of the world's greatest industrial nations. The development of transportation was the key to exploiting the hinterland, it seemed, and the chief means of transportation was the railway. The Canadian Pacific Railway, the Canadian Northern Railway, the National Transcontinental Railway, the Quebec Central, the Algoma Central were all expanding their systems of tracks. By the time the First World War was over, though, most of the railways were going bankrupt, and it was recognized that the railway system in Canada had been overbuilt. (It was this belated realization and the consequent governmental belt-tightening that killed all hopes for a railway to Red Lake.)

However, in 1900 "all systems were go" for the railroads. This was the year that the Ontario government, motivated by its desire to settle the northeastern part of the province, with its potentially good farmland in the clay belts, agreed to build a railway north from North Bay to New Liskeard. Construction of the 110-mile stretch began in 1902.

There is no doubt that the substantial silver find at Cobalt was made during the construction of this railway, then called the Temiskaming and Northern Ontario Railway. The discovery was made at Mile 103 on the railway, and the first fully-registered claim was staked in the fall of 1903 by Fred LaRose, a blacksmith working with the construction crews. The manner of his discovery, however, belongs in the realm of legend. The story goes that one evening in August "Lucky LaRose" was sitting by the campfire when he saw what seemed to be the eyes of a fox gleaming in the darkness. He threw his hammer, the hammer struck rock, and the "eyes" turned out to be nuggets of silver.

By 1905 a stampede was headed for Cobalt. This stampede, however, was in vivid contrast to the one to Red Lake 20 years later, as the prospectors rushing to Cobalt were able to go by train. Development of the silver veins was rapid. LaRose sold his find to the Timmins and McMartin brothers for $30,000.

Within ten years, Cobalt had 50 mines shipping ore and had produced $300,000,000 worth of metal.

The effects of the fabulous find at Cobalt, the most wealthy by far in Canadian history to date, were many and various. The first and most obvious effect was the immediate stimulus which the Cobalt find gave to prospecting in the area. This wave of prospecting activity had results just as fabulous as the finds at Cobalt. The gold discoveries at Porcupine, from 1905 on (with its Dome and Hollinger mines), and at Kirkland Lake, from 1909 (with Lake Shore, Teck-Hughes and Wright-Hargreaves), followed. The belt of gold mines extending from Timmins through South Porcupine and Kirkland Lake to Rouyn-Noranda has been called the "Canadian Rand", after the famous Rand properties in South Africa, which for many years was the only area in the world that produced more gold than the area in northeastern Ontario.

The railway followed these finds every step of the way. One train of note on the route to the mining fields was the T & NO's "Cobalt Special" that came into operation in 1907. This nightly train from Toronto, with its elegant library, luxurious dining facilities and Pullman cars, provided suitable accommodation for the brokers and financiers who financed the mines from offices in New York, Chicago, Montreal, and Toronto.

The second obvious effect of the Cobalt find came from the enormous amounts of capital it engendered. (In 1909 the *Cobalt Daily News* published a list of 35 millionaires whose fortunes had been made in the silver finds at Cobalt.) Money may breed contempt in some, but it also breeds more money, and the Cobalt discoveries sent waves of capital in all directions. The Timmins brothers, for example, who bought out Fred LaRose, used profits from that silver mine to buy and develop the enormously successful Hollinger property. Noah Timmins' interests later extended to the Noranda-Rouyn operation and the Iron Ore Company of Canada.

Similarly, the original Dome discoveries in the Porcupine area (so named because the original property included a fabulous underground "dome" of gold) had wide-reaching effects throughout Canada as the company expanded. Today the names of Dome Petroleum and Dome Mines appear daily in the news. And Dome has been of great significance in the history of the Red Lake area: first, negatively, when it bought and then dropped its option on the Howey property, causing the gold rush to abruptly fizzle out; and then, positively, when it bought the Campbell Red Lake property and brought into production, in 1949, the mine which has become one of the wealthiest in Canada.

The Cobalt-inspired mining boom caused significant economic growth, not only in mining, but in the economy in general. The city of Toronto gained enormously from it. It was the development of that large economic hinterland to the north, in fact, that enabled Toronto to outdistance Montreal as the commercial centre of Canada. Annual clearing house returns on the Toronto exchanges went from $800 million in 1902, to $5 billion after the discovery of Cobalt, and to $8 billion after Porcupine and Kirkland Lake got underway. Toronto had no rival as the centre for capital for mineral development; it became head-office territory, and it remained so when Red Lake came into production.

And from the Cobalt developments followed, not only other mining properties and important economic benefits, but a whole new crop of experienced prospectors, geologists and mining engineers. The first stage of the gold rush to Red Lake was dominated by experienced prospectors from the Porcupine area: the Howey brothers, Dan Willans, the McDonough brothers (who were everywhere in mining), the Cochenour brothers, and Alex Gillies (who had staked part of the Hollinger property). At a later stage in the development of Red Lake, Noah Timmins did $100,000 worth of development on the Cochenour property. And Harry Oakes, of the Lake Shore Mine, sat on the board of Howey Red Lake. There were other, less successful veterans of the Cobalt/Porcupine camps who found their way to Red Lake as well; people like Sandy McIntyre, who had first staked the land that became the very productive McIntyre mine and sold it for $125.00.

It has been said that the group that spent Christmas Day, 1925 together in Red Lake (just before the national and international press seized on the gold finds at Red Lake and the real rush began) was a "Who's Who" of Canadian prospecting. And every one of them had had experience, at one time or another, in the historically important developments in the Cobalt/ Porcupine region.

The process of turning a piece of virgin ground into a producing mine breaks down roughly into three stages. These steps might be called the prospecting, underground exploration, and surface development stages. This process has been the same, with certain technical changes, since the first gold mine was developed in Canada. The following outline, though, applies specifically to procedures followed in the 1920s.

The first stage began with the prospector. Working alone, or with a few friends, he arrived in an area which he had some reason to think had mineral potential. (The prospectors who went to Red Lake in 1925, and before, were not trained geologists, but they were experienced, knew how to recognize likely rock formations, and knew how to use professional geological material when it was available, such as maps or written analyses.)

Before the prospector could stake any claims, he had to have a Miner's License. This could be bought from any Mining Recorder's Office in Ontario and used anywhere in the province. There was no Mining Recorder's Office in Red Lake, though, until the gold rush was in full swing, so prospectors and would-be prospectors had to pay for and secure their license before entering the area.

With the license safely stowed away, and the likely piece of territory established, the prospector was ready to stake his claims. Each licensed miner was entitled to stake 3 claims for himself and 3 more for each of two absent licensees, for a total of 9 claims. In Ontario a claim could be 40 acres in area, that is, the area enclosed within a square having sides of 1320 feet, or 20 chains (a standard surveyor's measurement).

After determining the area he wanted his claim to cover, the staker would then stick posts, 4′ or 5′ long, in the ground at each of the 4 corners of the property. On the northeast post (the #1 post) he would write his name, license number, date and time. Then he would put the number of the post (2, 3, or 4) and his name on each of the other posts, going clockwise from the #1 post, around the property. He also had to blaze the trees on the perimeter of the square. A metal tag with the claim number on it had to be affixed to each of the posts; this could be done later.

After the claim was staked, the prospector then had to record it at the Mining Recorder's Office, for another small fee. (Unlike the miner's license, which could be bought at any Recorder's Office, the recording of the claim had to be done at the Recorder's Office for that particular Mining District. During the early part of the Red Lake rush, that meant another long trip out to Kenora after the claim was staked, just to record it.)

By staking and recording a claim, the prospector acquired the sole right to work on that piece of ground, or to sell it. To maintain that right, within the first 3 months, he had to perform 30 days of what was called "assessment work" on his claim; 40 days more in each of the next 4 years was also required; plus an additional 10 days, later on, to total 200 days of work within 5 years. He had one year beyond that 5 year period to pay all fees and get patent for the claim. (The assessment work required, usually consisted of making trenches on the property, sampling or panning, or just digging around to try to determine what was on the claim. A year's worth of assessment work, however, could be done in a week of diamond drilling.)

After the property had been prospected and staked, the second stage of the process, underground exploration, would begin with the assessment work itself. The point of this stage was to determine the size and shape of the underground ore body. Soil and vegetation (called "overburden") would be stripped away, trenches dug at intervals, and–if there were substantial funds around–diamond-drilling begun.

Diamond-drilling involved the drilling of deep holes in the rock, with power-driven drills that had diamond bits (of industrial-quality stones) at the tips. The long, narrow cylinders of rock extracted from the drilled holes (these cylinders were called "cores") would then be analysed, or assayed, to determine their mineral content.

If the trenching and diamond-drilling revealed that the property was of sufficient potential value (in gold-bearing properties, this evaluation was expressed in the number of ounces of gold, or the number of dollars, per ton of ore), then more extensive (and expensive) kinds of underground exploration would be carried out. This would involve sinking one or more shafts down into the rock, and then cutting out two or three horizontal levels from the shaft to get a more detailed picture of the ore-body. At this stage, more diamond drilling would be carried out.

If this stage, too, revealed good potential, the final stage of surface development would be undertaken. First, underground development would be brought to completion: the shaft would be sunk to a sufficient level, cross-cuts from it opened, and hoisting machinery installed to raise both the ore and the miners who would be drilling and blasting the ore out in chunks from the underground levels. Then surface facilities would be constructed to crush and mill the ore, in order to separate the gold from the rock (usually quartz) in which the gold veins were embedded.

This process was a complicated technical one which involved the importing of many pieces of machinery, and the construction of buildings to house them. In the process itself, large pieces of ore were first given a preliminary sorting and then crushed to a fine pulp. The larger particles of gold in this pulp could be freed relatively easily by passing the pulp over plates treated with mercury; the gold amalgamated to the mercury, the waste passed over, and then the gold and mercury were separated in a retort furnace.

That waste pulp from the amalgamation process still contained a large percentage of gold. To separate these fine particles of gold from the worthless rock, the pulp was then immersed in a cyanide solution. The presence of the cyanide caused the gold to separate out from, or dissolve out of, the surrounding rock; the gold was then precipitated (condensed out of the cyanide solution as a solid) by contact with zinc, to which it adhered. It was then separated from the zinc, refined in a blast furnace and poured into brick moulds to be shipped away.

Along with the construction of buildings to house the machinery, all the necessary support structures would be erected: office buildings, a cookery, bunkhouses for the single men, perhaps a store and a townsite of homes for families.

Certain of the last elements in the third stage of mine development were optional; some mines in some areas had no townsites of their own, and occasionally an ore-body that was in very close proximity to an existing mine might have its ore processed by using the older mine's milling facilities. But the general pattern was essential to every gold mining operation, because of the type of ore involved. Because that ore was always relatively low grade, in terms of ounces of gold (or dollars of profit) per ton of ore, the ore had to be processed on the spot rather than shipped out, at great expense, for processing elsewhere. None of the major gold mines of Ontario have had nuggets available for the plucking like those that gleamed up at the lucky Klondikers from the bed of Bonanza Creek. The gold of Ontario had the odd surface showing, but the mass of it was found in veins meandering through underground rock.

The kind of ore, then, dictated both the timing and the type of development which were possible for an area like Red Lake. When the railway had established the basic lines of communication and transportation across Canada at the end of the last century, it also dictated the areas of settlement. Settlement, and development of any kind, whether agricultural, timber, or mineral, occurred right alongside the railway, or else in small circles centred around it. The 180 miles of lake, rock and forest that separated Red Lake from the railway was too great a distance over which to bring all the heavy, awkward and expensive machinery that a lode gold mine had to have.

As a result, it was not until the gasoline engine became sufficiently sophisticated and cheap to operate that such an area could be developed. (And that is why the gold rush and ensuing development in Red Lake depended so heavily on both the primitive methods of transportation like the canoe and the dog team, and on more modern forms like tractors, motor launches, scows, and, most radically, the airplane.)

The stubborn ore also determined just how developments were to be financed. Lode gold mining was (and is) an expensive proposition, and it was a very rare property that could be developed by a shoestring operation. The Lake Shore Mine at Kirkland Lake is the only lode gold mine in Canadian history which was staked, developed, and brought into production by one man. And that man, Harry Oakes (later Sir Harry) became one of the wealthiest men in Canada. Such large personal fortunes in lode gold mining, however, were rare. This is not to deny that certain stakers made handy sums of money when they sold their properties, or that senior executives (who were often major shareholders as well) of successful producing mines enjoyed six-figure incomes for many years.

The most common method of financing a gold mine during the early part of this century (and the method used to bring the first gold mine at Red Lake, the Howey, into production) broke down, roughly, into three stages. In the first stage, the prospector's initial explorations were usually financed by a small "syndicate", a group of friends, neighbours, relatives, who would each contribute a modest sum to support the prospector as he covered the ground in some likely area. In exchange for their financial support, they would have signed an agreement to give them a certain share of ownership in any properties staked.

Should the prospector find a property which seemed worthwhile, he, with the syndicate backing him, would then set out to sell their rights in that property, on the best possible terms, to someone who had the large amounts of capital necessary to do a thorough examination and exploration of the property. In this second stage of financing, the prospector and his friends would probably approach a wealthy speculator, or, more likely, a mining broker.

The broker, after making sure that the mining property in question had the potential for sufficient profit to make his interest worthwhile, would then form a corporation. A certain percentage (perhaps a third) of the shares he would reserve for himself, and the rest would be offered to the public at a relatively low price. The money gained from this initial sale would be used for diamond drilling, to further pinpoint the value of the mineral deposit. At this stage as well, an option on the property might be taken up by a large mining company who wanted to expand its holdings. This meant that the large corporation would buy the right to acquire a certain percentage of the stock, after a specified amount of time had elapsed, should it choose to do so. In the intervening time, the corporation would pay for this right by doing an agreed amount of exploratory work (usually diamond drilling) on the property.

In almost all cases, such options were dropped. In part, of course, because the properties were not promising enough, but also because the large corporations were very conservative in nature. If this option were dropped, that was usually the end of the property's development. Similarly, the end might have come during the second stage, before any options had been taken up, as diamond drilling, financed by the sale of penny stock, might have failed to show enough promise.

But very occasionally, even though the options were dropped, the broker or speculator was able to find other sources of development capital, and the third stage of financing was reached. (This stage could also be attained though no option had ever occurred, if the penny-stock-funded initial diamond drilling had shown good results.) This was the stage which resulted in a producing mine. More stock would be offered for sale, at higher prices, and the money from this would be used to sink a shaft. If further explorations of the shaft continued to show profit potential, a mill and other necessary buildings would be constructed on the site.

During the 1920s there was a mining boom in Canada that coincided with a boom period in the economy as a whole. The wealthy mines in northeastern Ontario were at peak production and everyone was hoping for new discoveries that would be just as rich. A whole mining-centred financial system had developed in Toronto. The number of brokerage houses specializing in mining stocks was rising rapidly, as was the price of seats on the Mining Exchange (The Standard Stock and Mining Exchange remained separate from the Toronto Stock Exchange until 1934.) The public, who delighted in speculating at this time, was particularly keen on the penny mining stocks.

The intense interest in mining had several results as far as Red Lake was concerned. It was probably responsible for the Howeys' discovery of gold at Red Lake, in the first place, and for the fact that Jack Hammell was eventually able to get sufficient funds to bring their property into production. (And the Howey Mine itself was the fountainhead of all the subsequent development at Red Lake.)

It was also responsible for the worldwide intensive press coverage of the Howey find, which both resulted from and helped to cause the Red Lake gold rush. But the whole feverish, speculative attitude of the time also caused the startling collapse of the gold rush. Of the approximately 6,000 claims in the Patricia District staked during the gold rush, only some 50 properties were actively examined. Of these, only one at Red Lake, the Howey, came into production in the first ten years after the Howey brothers' initial discovery.

The Gold Rush, Part I: July-December, 1925

The sequence of events leading to the Howey gold discovery that opened up the Red Lake area, began in Toronto. A veteran prospector named Findley McCallum, who had staked claims in the Red Lake area during the short-lived silver excitement of 1922, had seen a copy of the Ontario Department of Mines Report on the Red Lake area written by Dr. Bruce. It seemed to McCallum that this report, published in 1924, was encouraging enough to bring in to Red Lake some of the development capital which had, so far, steadfastly refused to get involved there.

In the spring of 1925 McCallum went to Toronto, the source of most mining capital, to try his luck. As far as we know, he was unable to find any backers, but he did meet a fellow prospector, in the King Edward Hotel, with whom he shared his story. This fellow prospector was Lorne Howey.

Lorne and his brother Ray were relatively young specimens of that now almost-extinct species, the inveterate prospector. They were living in Haileybury in 1925, and their search for mineral wealth had taken them, in the past, to Porcupine, to Gowganda, to Lightning River, and to northern Quebec. Lorne was described at the time as a man with "keen blue eyes, sober, weather-beaten face and active bearing. . . . At 41, with half his life spent in the quest of minerals, he looks 30 and no-one would take him for anything but what he is, a prospector."[17]

When Lorne was a boy in Guelph he wanted to be a newspaper reporter. He had intended to look for a job with the *Toronto Star*, but when the proprietor of the *New Liskeard Herald* offered him a job there, he took it. He left that job after a few years for the more adventurous and independent life of a prospector, and had been 20 years at that when he went to Red Lake and made his find. A newspaper reporter later asked him if he attributed his find at Red Lake to luck. "Luck . . . well, maybe, but luck is not much good if you trust to it. There's a good deal in working along a systematic plan, keeping your

----

[17] *Toronto Star,* February 16, 1926.

head, and sticking at it."[18]

True words, certainly, but nonetheless Lorne Howey's encounter with Findley McCallum does have an aura of good fortune about it. McCallum showed Howey his copy of Dr. Bruce's report, apparently, and Howey went over to the Department of Mines office to get a copy of his own. Armed with the report, he returned to Haileybury and set about finding enough money to support him for a summer's prospecting (in prospector's terms, to "grubstake" him). He and his partner, his brother-in-law George McNeely, succeeded in getting a group of Haileybury citizens to fund them. (This was the first Howey syndicate.) Lorne's brother Ray, who was prospecting at the time for the large and wealthy McIntyre-Porcupine Mines, decided to accompany Lorne to Red Lake. He took W. F. Morgan, another McIntyre-Porcupine employee, along with him.

The two parties, travelling together, reached Red Lake in early June and discovered four prospectors from the Haileybury area already there. Bobby Cockeram and Russ Ferguson, old buddies of Lorne Howey's in the loose and comradely mining fraternity, were scouting the shores of Red Lake, and so were two prospecting mining engineers who worked for Dome Mines, Alex Robertson and Jim McCrea.

It seems they all prospected together, unsuccessfully, for a month or so, investigating the shores of the lake. Apparently the McCrea and Cockeram parties decided that the Woman Lake area, about 75 miles east of Red Lake by water (and about 50 miles by air) seemed a more likely place to prospect, and they left accordingly.

The two Howeys, McNeely and Morgan remained camped in what was called "Burnt Bay" (what is now Howey Bay), because Lorne had cut his foot with an axe, and because a strong north wind made it seem advisable to wait. Although there are two differing stories concerning their find of free gold (visible gold, that could be easily flaked or chipped off), the date is known: July 25, 1925. According to one story, it was while hunting deer close to camp that they came upon a place where

[18] *Toronto Star*, February 16, 1926.

a large tree had fallen. Near its roots, the earth had been torn away from the surrounding rock, revealing small veins containing free gold. According to the other story, it was while following a likely-looking quartz vein that gold was found, first by Lorne and then Ray.

Whatever the immediate circumstances, they were elated. "Looks like we found it!" exclaimed Lorne. And the more stoical Ray is said to have replied, "Ain't that what we came for?"

It was obvious to Lorne and Ray that their virtually simultaneous finds were outcroppings of the same ore-body. Since they were prospecting for different backers, a rough-and-ready, fair division was obtained by pacing off the distance between the two discoveries, and drawing an east-west line at the halfway point. Ray Howey and Morgan then staked 9 claims south of this line for McIntyre-Porcupine, while Lorne Howey and McNeely staked 13 claims north of the line for themselves and their backers in Haileybury. (Thirteen was, Lorne explained later in a newspaper interview, his lucky number.)[19]

The group left the Red Lake area around the beginning of September and returned to Haileybury. (As Lorne said, "Then we went out to record our find and tell our friends.")[20] From that point on, things moved very quickly. The first necessity for Howey and McNeely and for the small group of grub-stakers who had financed their prospecting, was to find further financing for exploration of their claims. To this end they contacted Jack Hammell.

Hammell was a well-known mining promoter in Toronto. He had started his working life as a busboy in Toronto and first began to deal in mining claims while a bartender in Cobalt. One of his greatest successes in mining was the discovery of the large copper-gold-zinc deposits at Flin Flon, in northern Manitoba, for which he had just arranged development financing in 1925. He was eventually responsible for the development of a number of mines in northern Ontario, including the Pickle Crow

----

[19] *Fort William Times-Journal,* March 15, 1926.

[20] *Fort William Times-Journal,* March 15, 1926.

and Uchi mines and, in the Red Lake area, the Hasaga and the Howey.

Over the years Hammell was responsible for raising millions of dollars for the development of Canadian mining. He was also a major figure in promoting northern flying through his N.A.M.E. program (Northern Aerial Minerals Exploration) whose pilots covered more than a million miles in the exploration and development of mining projects. Hammell was responsible for the development of the Red Lake area, not only through his financing of the Howey, but through his subsequent relentless pressure on both the Ontario and the federal governments to improve transportation to Red Lake and provide services within the town. He seems to have had, with much justification, a great deal of confidence in his abilities as a mover and a shaker: Harold Farrington, who had frequent contacts with him, said that Hammell used to sing his own variation on a popular song: "The whole world is waiting for the sunrise . . . and Jack Hammell."

When Lorne Howey approached Hammell in the late summer of 1925, Hammell was interested in the description and the samples, but wanted to see the claims before committing himself to anything concrete. He made a quick trip in to Red Lake, accompanied by Cooney Woods and Alex Gillies. They arrived at Burnt/Howey Bay on September 25, and did a fast inspection of the properties. Gillies, who had been a scout for McIntyre-Porcupine for three years, was much impressed. And so, obviously, was Hammell, as he proceeded to stake further claims adjoining the Howey-McNeely ones.

Hammell was not the only man with inside information who was staking claims in Red Lake at that point. Everyone concerned had been very careful to keep the news of the discovery of free gold as quiet as possible, but, nevertheless, two other parties from the Haileybury region also went to Red Lake in September. (Perhaps because they were old prospecting buddies of Lorne Howey's, he had told them–and perhaps not.) In any case Bill and Ed Cochenour and Dan Willans staked 18 claims adjoining the Howey and McIntyre claims to the south and west. These became the Buffalo Red Lake Syndicate. They also staked, on their next trip to Red Lake, the property which was eventually to become a solid, producing mine, the Cochenour-Willans. The second group of prospectors, Bobby Cockeram, Russ Ferguson and Bill Smith (the first two of these were back to Red Lake for the second time that season) staked 12 claims, which became the Cockeram Syndicate.

Jack Hammell, the professional entrepreneur of the lot, had no time to waste, and was on his way back out again on the stormy fall lakes before the end of September. On his way out he met George Swain, the medical-student-turned-woodsman who had operated a trading post at Gullrock Lake, just south of Red Lake on the Chukuni River, since 1918.

Swain was in the process of moving his four tons of trading goods and supplies, singlehandedly, over to Woman Lake in search of greener trading fields. Apparently Hammell told Swain that free gold had been discovered at Red Lake and that Swain should go and stake some claims for himself. Swain refused, pointing to his four tons of goods laying on the rocks of the portage where the two men had met. He was not interested in staking claims; he was not interested in moving his post to Red Lake in the hopes of serving a possible influx of prospectors; he was going to Woman Lake–and that is where he went. Unlike the thousands of greenhorns who mushed in to Red Lake to lay worthless claims on the snow several months later, Swain might indeed have been lucky had he followed Hammell's advice.

As soon as Hammell got back to Toronto, he made an agreement with McIntyre-Porcupine, owners of the claims staked by Ray Howey, to do initial development work on their property. At the same time, he organized a new company to develop the property Lorne Howey had staked. When Lorne Howey formed his original prospector's syndicate, to finance his prospecting at Red Lake, he had sold shares in it for $25.00 each. The new company that Hammell formed, the Howey Gold Mine Syndicate, consisted of 5,000 units: half of these were owned by Hammell, and half by the stakers and their original prospector's syndicate. (When the conversion from prospector's syndicate to Hammell's syndicate was made, one share in the first was worth 77 units in the second.) At the time that Hammell created the Howey Gold Mine Syndicate, he also agreed to spend $50,000 developing the property.

The units in the Howey Gold Mine Syndicate, which existed until Howey Gold Mines was incorporated in late 1926, rose rapidly in value on the mining exchange after the gold rush to Red Lake got underway. At one point a unit was selling for $700.00. A hypothetical backer of Lorne Howey, who had bought one $25.00 share in his prospector's syndicate, would therefore have been able to make the handy sum of $53,900 (or, $700 x 77) had he sold when the units were at their highest price. Later, when the Howey Gold Mines operation was constituted, units in the second syndicate were exchanged for 400 shares each in the new corporation.

Meanwhile, back at Red Lake, the Cochenour and Cockeram parties stayed on, prospecting and staking, until mid-October. That was the latest point in the season that one could easily move in and out of the area, as ice usually began to form in the bays. There were a total of about two dozen claims staked (many of them by proxy) at Red Lake during the autumn, after the initial Howey staking. These were staked by Hammell; by the Cockeram and Cochenour parties; by a couple of employees of the Hudson's Bay Company; and by Major C.J.A. Cunningham-Dunlop, Chief Exploration Engineer for the Coniagas Mines at Cobalt. Cunningham-Dunlop was a man of enterprise, who, on returning east, immediately resigned from the employment of Coniagas and formed his own syndicate. He was also a man of many words, as he gave frequent and lengthy interviews to the press later on, interviews which helped feed the flames of excitement about this new gold discovery in the romantic north.

All the claims registered in the late summer and autumn of 1925, with the single exception of Hammell's, were for residents of northern Ontario: in the files of the Mining Register's Office their addresses were listed as Coniston, Cobalt, Haileybury, Sault Ste. Marie, Fort William, Hudson, and South Porcupine. These claims were prefixed with the letters KRL because at that point Red Lake was still in the Kenora Mining Division; this meant that the Mining Recorder's Office for the area that included Red Lake was in Kenora, and, therefore, that every claim staked in the Red Lake area had to be recorded there.

To the veteran prospectors and other northern residents who staked claims during that period, the lengthy and difficult route from the railway by water, into Red Lake and then back out to Kenora, was hard work, certainly, but nothing out of the ordinary. They carried with them only the tools of their trade–though of course those tools, food, tent and canoe made a substantial enough load over the portages.

The route most of them followed into Red Lake was one of the routes that had been used by the fur-traders (and before their advent, by the Indians). The distance from the railhead to Red Lake was about 180 miles by water, in summer, and about 40 miles less in winter, when a more straightforward route was possible. The men who came in to Red Lake during the early part of the gold rush usually disembarked from the train at Quibell, on the CN line, whereas the later jumping-off point (and the one used by the large numbers who flocked in after Christmas) was at Hudson, to the east. Starting the journey at Hudson made for a trail that was longer but easier to travel.

In either case, the large body of water known as Lac Seul covered more than half the distance from the railhead to Red Lake. Power boats could travel only this distance. After one reached the north-west end of Lac Seul, where the Hudson's Bay Company's Pine Ridge post was located, the portages began. There were some 5 or 6 in number, depending on the exact route chosen. The trail went west from Pine Ridge along the English River (here, there were two short portages or one longer one) to Pakwash Lake, then from Pakwash Lake to Gullrock Lake along the Chukuni River (there were four short portages here) and from there along the Chukuni to Red Lake. The trip by canoe from the railway to Red Lake, during reasonable weather, took 5 or 6 days.

At this time, there were steam- and gasoline-powered launches traversing Lac Seul with some regularity during the summer months. The Bowman Fish Company, for example, had a launch called the Victor which made a regular run to Goose Island, a short distance from Pine Ridge. And the Hudson's Bay

Company had been operating its own launch–first steam-powered, then gasoline-powered, when that proved cheaper–since 1915. During the winter season, on the other hand, no transportation was available but dog team.

There is no evidence that these fast and modern modes of transport were used by the stakers of 1925. The launches in question were not passenger boats, their routes and times were not flexible, and of course a canoe propelled by man-power was a much cheaper mode of transportation. The only known exception, at that point, to this reliance on the canoe occurred in October. Two surveyors working for the Dominion Land Survey, McCusker and his assistant Malaher by name, had been working in the Berens River area, gathering surveying data which was to be worked in with that provided by aerial photographs, being taken by the Royal Canadian Air Force.

McCusker and Malaher arrived in Red Lake from the north on October 3rd, and apparently discovered the location of the Hudson's Bay post by following the sound of howling sleigh-dogs. (This was just a modest foretaste of the din that filled the cold clear air around Howey Bay several months later, with thousands of sleigh-dogs howling in disharmony.) When the two surveyors arrived at the post, the manager William Smith told them of the exciting gold discoveries in the area. Smith and Malaher planned to stake some claims the next day. (McCusker, because of his position, was unable to do so.) It is not known whether they did so at that time or not; no claim was recorded in Malaher's name, but one is recorded in the name of William Smith for October 23.

The weather, however, began to change, at that point, for the worse. Rain and snow began to fall. The surveyors did not want to be locked in by ice, and they made fast time, by canoe, out to Pine Ridge where they caught a ride on the Victor back to the railway.

Neither canoe nor fishing-launch, however, was sufficient for Jack Hammell's purposes. He was not the only man in Red Lake, that fall, who knew that it took a lot of capital and an equal amount of entrepreneurial know-how to develop a producing mine from a piece of virgin ground; but he was the only man there, at that time, who had the capital, the instinct and the connections to do so.

It was October, freeze-up was fast approaching, and Hammell needed to bring in men and equipment to do the initial surface exploration on his claims before winter set in. To do this, he made a decision which turned out to be a revolutionary one: he turned to air transport. This was the first time planes had been used to open up new territory for any kind of development, and it proved to be the beginning of a long and fruitful association between the airplane and the development of northern territory. Hammell's decision also made the Red Lake gold rush the last one ever to rely on the primitive modes of canoe, snowshoe, toboggan and dogsled, and it made the rush itself a bizarre combination of the most modern and the most ancient methods of transportation.

The only planes (and the only air service) within striking distance of Red Lake, in the fall of 1925, were operated by a branch of the Ontario Department of Lands and Forests, called the Ontario Provincial Air Service. This service owned seven Curtiss HS2L Flying Boats, based at Sioux Lookout. These unwieldy-looking crafts had been designed and built during the First World War to patrol coastal areas, and were capable of carrying a total load (people and freight) of 650 pounds. Hammell chartered five of the Flying Boats. They were flown west to Minaki, which was appreciably closer to Red Lake by air than Sioux Lookout, and was also on the CN line, so that crews and supplies could go at least that far by rail.

The five aircraft went back and forth from Minaki to Red Lake, carrying a total of some 15 tons of supplies and a seven-person crew, until final freeze-up at the end of October made the waters of Howey Bay inoperable for them. One of the pilots flying into Red Lake at that point was Harold or "Doc" Oaks, a mining-engineer-turned-pilot who became one of the foremost figures in Canadian aviation.

The crew that was flown in to Red Lake in October to do preliminary work on the Howey property included one woman, the first white woman known to have lived at Red Lake. Her name was Mrs. Palo (her first name is unknown), and she was described by a newspaper reporter as large, blonde and smiling. She and her husband were both Finns. Her husband, a sailor who had drifted to many parts of the world, found himself in the unlikely spot of Sioux Lookout in 1924, working as a millwright, and Mrs. Palo came from Helsinki to join him.

Things did not work out well for them: they had a son, who died, and Palo couldn't find a steady job. In the fall of 1925, Palo met Alex Gillies at Sioux Lookout, and Gillies offered him a job doing trenching on the Howey property at Red Lake. Mrs. Palo didn't want to be left alone, so arrangements were made for her to be hired as cook in Red Lake as well–and Palo's job changed to that of her assistant.

The basic crew consisted of seven men, but others came and went, and Mrs. Palo cooked for as many as 35 at one time during her stint at Red Lake. She would arise at 3:30 a.m. to begin baking the day's bread, and usually finished around 9 in the evening. She spent the whole winter following this regime in the Howey tent. There were other Finns in the crew, and the group built a bath-house, one of the few log buildings in the group of tents that housed the Howey camp. She went back to Sioux Lookout with her husband and several crew members in March of 1926. Their method of transportation was a unique one for the Red Lake trail, as they went on skis. After the trip out from Red Lake, Mrs. Palo drops from view.

When the initial crew was at Red Lake, Alex Gillies was in charge of the operation there. Gillies, who had first come to Red Lake in September with Jack Hammell to check out the McIntyre property, was another veteran of the Cobalt/Haileybury school, as he had been one of the original stakers of the Hollinger Mine in 1909. Gillies' network of old mining buddies included Gus McManus. McManus had staked claims on the shore of Red Lake near the mouth of the Chukuni River in 1922; he had found some gold on his claims, but hadn't been able to arouse any interest in his property.

When Gillies was in Haileybury at the end of September (after his quick trip in to Red Lake with Hammell, and before he went back in with a crew to start work) he told his friend Gus to go back to Red Lake and restake his old claims. Gillies told him that the best way in was from Quibell, on the CN line, and that he should get moving quickly.

McManus took another mining veteran, Jack St. Paul of South Porcupine, as his partner, and they made their plans, got their dog team and supplies together, and set off by train. It was the first day of December when they reached the point on the CN line where they had decided to disembark for Red Lake, and by now news of the Howey find at Red Lake had begun to seep out beyond the tight-knit Haileybury circle. McManus knew that several other parties of prospectors were on his trail, and decided to make that trail as difficult as possible to follow. (McManus and St. Paul, though they hadn't taken that particular trail to Red Lake before, were nevertheless far better informed about the location of routes to Red Lake than the others.)

McManus and St. Paul, in fact, departed from Favel Station, 12 miles to the west, rather than Quibell. They took 18 days to make the journey from the railway to Red Lake: in part because the going was slow, over new soft snow, and in part because they made a number of false trails to divert the prospectors behind them. Their tactics were successful, as the McManus party left the railhead one day before their eager followers and arrived at Red Lake three days in advance of them.

When McManus and St. Paul arrived at Red Lake on the 19th of December, they visited the Gillies camp and turned down offers of hospitality in order to get at their staking–but they did accept an invitation to Christmas dinner. For several days they worked their way around the lake and finished by restaking the original McManus claims. According to McManus, his race with the prospectors behind him ended dramatically. Just as he was staking his last claim he heard sleigh dogs barking. "Here they come!" he called to his partner, and around the corner came two dog teams. "Merry Christmas!" shouted the new arrivals.

McManus then realized it was Christmas Day, so he and St. Paul hastened over to the Gillies tent for Christmas dinner. The gold rush had begun.

The Gold Rush, Part II: January-February 1926

Christmas of 1925 provides a convenient point for dividing the Red Lake gold rush into two parts. The first stage lasted from the Howey discovery, at the end of July, until Christmas, and during this stage it was almost exclusively men from the Cobalt/ Haileybury/Porcupine area, an inner circle of seasoned prospectors and other mining men, who heard of the Howey brothers' discovery and went to Red Lake on the strength of what they heard to stake claims for themselves.

The second stage lasted for a year. It began at Christmas of 1925 when men from outside the small mining fraternity first heard the news. The gold rush to Red Lake, measured by the number of hopefuls who thronged in from Hudson, reached its peak in the late winter-early spring of 1926. And it was over, absolutely, by Christmas of 1926, when the dozen or so die-hard prospectors celebrating Christmas on the shores of Howey Bay formed a forlorn group of almost exactly the same size as the optimistic one that had gathered there a year before.

On the other hand, if the dividing point in the gold rush is extended from Christmas to the end of January, it creates another productive point of view for looking at the gold rush. For by the end of January–that is, before the thousands of greenhorns from many parts of the world began to rush across the snow and ice from Hudson to make their fortunes–all the good land was already staked. In fact, with the single exception of Campbell Red Lake, the property of every significant producing mine in the Red Lake area was staked by the end of January, 1926.

Excluding Campbell for the moment (although parts of its present property were staked by Major Cunningham-Dunlop in early 1926), the Red Lake district has produced virtually all its gold on land staked before the real gold rush began. The Howey Mine, whose success was ultimately responsible for the development of all the other properties in the area, resulted from Lorne Howey's initial staking in July, and the Hasaga mine from his brother Ray's. (These mines both produced moderate amounts of gold: the Howey, just over $13 million worth, and the Hasaga, just over $8 million. Another moderate producer, Starratt-Olsen, first staked in January of 1926, produced $6 million.)

McKenzie Red Lake, first staked in 1922, was restaked in December of 1925 by William McNeely, and eventually produced $24 million in gold bullion. Cochenour-Willans, most of whose claims were staked in January of 1926 by Bill Cochenour and Dan Willans, produced over $41 million. Madsen Red Lake, also first staked in January, produced close to $82 million. Dickenson, also staked in January of 1926, has been producing gold steadily from 1948 to the present.

Three small producers also date from this time: Red Lake Gold Shore, which produced $750,000, was staked in January, 1926; Gold Eagle, which produced $1,500,000, was first staked in 1922 but the claims changed hands in early 1926; McMarmac Red Lake, staked in December of 1925, produced about $3 million. In every case it took some years, and often several changes of ownership of the claims, before the property was developed into a producing mine. (The longest such interval occurred with Dickenson, which did not begin production until 1948.)

January 1926

The last days of 1925, then, saw a number of prospecting parties set out on the trail to Red Lake. One of the largest of these parties was led by Major Cunningham-Dunlop, recently resigned as Chief Explorations Engineer for the Coniagas Mine in Cobalt. After his initial trip in to Red Lake to look around and stake for Coniagas in the autumn, Dunlop was sufficiently impressed by the look of things, not only to resign from Coniagas, but also to drum up enough support from would-be financiers to form, not one, but three new mining syndicates: the Dunlop Red Lake, the Toronto Red Lake, and the Patricia Exploration Syndicates. Cunningham-Dunlop, a member of a prominent family in Canadian mining, died in 1932, before his claims were developed. Had he lived longer, he might have become a very wealthy man.

The Dunlop party that disembarked from Hudson in the last days of 1925 must have been an impressive sight in that straggling hamlet, as it was yet unused to the crowds and din that was to follow. Dunlop was head of a party of seven and was accompanied by five dog teams of eight dogs each. After his party reached Red Lake, they stayed there for about three weeks, staking a total of 50 claims in what is now Balmer Township and covers what later became the Dickenson Mine, and part of Campbell Red Lake.

When Dunlop came back out, on the trail, to Hudson and from there back to Toronto, his descriptions of his experiences at Red Lake became the first in the eastern press to have any personal details about the new, exciting developments at Red Lake. There had been brief notices concerning developments at Red Lake in the Manitoba press since early January, though, and on January 26 both the *Winnipeg Tribune* and the *Manitoba Free Press* quoted W. A. Kingsland, General Manager of the CNR, to the effect that prospectors and dog teams had begun to pour into Hudson and a gold rush to Red Lake had begun. Similarly, the *Toronto Star* had announced on January 8 that "Red Lake Situation Looks Interesting" and referred to the "long, expensive and hazardous trip" there.

The *Manitoba Free Press* sent a reporter on the trail to Red Lake at the end of January, and he reported that some 200 "prospectors, miners, and employees of commercial companies" were wintering there, with expectations that the number would increase to 1,000 by spring.

Major Cunningham-Dunlop's descriptions of his January trip to Red Lake, however, had a more romantic flavour to them. The *Toronto Mail & Empire's* account of a conversation with Dunlop might well come from a movie script. "All the scenes of romance are there, with the dogtrain, the pack and the packer, the snowshoes, the Ojibwa Indians, and the general excitement which always accompanies the thought of gold." The numbers of people at Hudson when Dunlop arrived back there on his way out at the end of January were increasing rapidly, and he announced to the *Mail & Empire* that "Hundreds of pack-trains are being made up, and they are trekking out of Hudson at the rate of from ten to twenty a day."

Dunlop's description to the reporter of Red Lake itself was interesting. "And, after one's arrival at the hub of the gold rush, there is little more of civilization than he found along the way. For the Red Lake District is so wild and remote from the ordinary routes of man that the territory has not yet been surveyed, and the only habitation in the whole district is the Hudson Bay Post. . . . The man who finds himself at Red Lake might well imagine that he was back in the days of early Canada."[21]

And to the *Toronto Star,* Dunlop is reported to have pronounced that there had been no such rush to a new district since the Klondike. (And this remark was made well before the rush began in earnest.) It was heady statements such as these, concocted in small part by men like Dunlop who had been in Red Lake, and in large part by an eager press, that contributed to the gold rush itself. As the statements about Red Lake's mineral wealth grew more lavish, they were picked up by newspapers around the world, and read by many hopeful novices who decided that they too would "mush in" to try their luck.

In January of 1926, though, the land described so optimistically by Major Dunlop was still only slowly being staked. The large Dunlop party had been followed out of Hudson, a few days after their departure for Red Lake, by two engineers for Dome Mines, Jim McCrea and Alex Robertson. There had been a fresh fall of snow since the Dunlop party left, and the two Dome men had to break trail all the way, which made for a slow, hard trip of 14 days. (The average time on a good winter trail was about a week, but a particularly hardy man like Neh Faulkenham could do it in four days.) When McCrea and Robertson arrived at Red Lake they staked claims for Dome Mines, just north of the Howey claims.

The Dome party was followed in to Red Lake by a group of three prospectors who had been sent by a speedily organized syndicate of Cobalt businessmen. These three veterans of the north (Mungo Williamson, Joseph Sweet and Bill Stadelman by name) followed, intentionally, a day behind the Dome party to take advantage of an already broken trail. It seems they were

----

[21] Quoted in *Canadian Mining Journal,* February 19, 1926.

not only wily but lucky, as the three groups of claims which they staked in the Chukuni River area, were later sold for $15,000 each.

More prospectors were arriving. Jack St. Paul, who had come in with Gus McManus in December, had wired his brother Cyr to come in quickly and stake some claims. Cyr and a partner did so. They found the ground directly west of the Howey claims still unstaked, presumably because it had been identified, on a geological map, as pure granite and therefore unlikely to carry gold. Nevertheless, they staked claims there. Two other prospectors from Massey arrived a few days later. They were being grub-staked by a Toronto syndicate, headed by a Dr. Bruels. This pair of prospectors, of whom Jack Campbell was one, reasoned that the Howey would one day be a big mine and, therefore, that all the land around it would one day be valuable real estate, so they staked 16 claims there too. Both groups received a pleasant surprise when the snow melted and they discovered that their "granite" claims contained quartz veins with good gold values. These 22 claims were later amalgamated as the Red Lake Gold Shore Mine, with Dr. Bruels as President.

The property which became the Cochenour-Willans mine was staked by another group of prospectors from the Haileybury area, Bill and Ed Cochenour and Dan Willans. The Cochenours were friends of Lorne Howey, and as soon as they heard of his discovery at Red Lake, they immediately took the trail there. Bill went in September, and staked some claims to the south and west of the Howey ones. He returned with his brother and Dan Willans in January, and then staked the claims which proved to be the good ones, some miles to the north and on the east shore of Red Lake, opposite McKenzie Island.

Bill and Ed Cochenour were Haileybury boys, born and bred to the north, but their partner was somewhat of a mystery man. Dan Willans had appeared out of nowhere in Haileybury in 1908, and never spoke of his background. Unlike most other prospectors, he seemed to be, not only a cultured gentleman, but a British one. Like many other prospectors, though, whenever he had sold some mining claims he went off to the city (Toronto, in his case) and lived the high life off the proceeds for a few weeks.

He and the Cochenours found it difficult to find capital for the property which they had staked, and it was not until 1936 that Cochenour-Willans Gold Mines was incorporated. In the fall of that year, Willans set off by canoe from Iroquois Falls to work on some claims he held in that area, and he disappeared without a trace. (A friend speculated that he had committed suicide, for Willans had recently told him that he was getting too old for the bush, and reaching the end of his tether as a prospector.) Many men have disappeared on the water, but very few have left fortunes without an heir. After the mine began producing in 1939, a sizeable sum of money accumulated in Willans' name. Advertisement was made in several countries for heirs to this fortune, but no-one ever presented himself to claim it, and the money eventually went to the government.

An adventurer who later became an important figure in mining at Red Lake also disembarked from the CN line at Hudson in January to head north for gold. On January 12 a young Dane named Marius Madsen and his party of five headed north from Hudson by dogsled.

Madsen, who was 25 at the time, had already had one great adventure. At the age of 19 he had been hired as part of the small crew of the Dagny, a ship owned by the Danish East Greenland Company. All crew members were trained to carry out scientific observations on the grim and inhospitable east coast of Greenland. Before the ship reached its destination, though, it was crushed by ice. Madsen and 8 other survivors lived for a year on a small island near the coast, and when the rescue ship arrived, Madsen elected to stay on for another year to carry out the work he had originally been hired to do. The battle for survival he had won in that extremely cold and barren region made the rigours of the Red Lake trail negligible by comparison.

After the Madsen party reached the Howey claims in January, 1926, they proceeded 20 miles west to Trout Bay, at the west end of the lake. There, the group staked a large area of 45 claims. These claims, like the majority of those staked during the gold rush, were never worked, however.

Madsen was interviewed by a newspaper reporter at Hudson on his way back out, and said that the trail at that time was an almost continuous stream of men, dogs, and transportation devices of every kind. He also said that the trail was littered with the carcasses of dead dogs, with one lying every hundred feet along the way. (This unpleasant aspect was not noted by any other participant–and was later denied by several.)

When Madsen headed back to Hudson by dog team in early March, though, it was not the end of the Red Lake trail for him. For the next eight years Madsen continued to prospect around the Red Lake area, and to study all the technical information on gold formations that he could lay his hands on. In 1934 he became interested in the area near Faulkenham, Flat, and Russett Lakes, largely because of the geological features of that area as presented on Bruce's map. Much of this ground had been staked twice by then–first by Neh Faulkenham in January, 1926, and then in 1933 by another group of prospectors–but all claims had lapsed.

In 1934 and 1935 Madsen secured much of that land, and attempted to find $15,000 from development companies and financiers to do initial development work on the property. He eventually made a $91,000 deal with Joe McDonough, a mining veteran who had gone to Red Lake in January of 1926 as well. McDonough became the first President of Madsen Red Lake Gold Mines Limited, which was incorporated in March of 1935 and produced almost $82 million in gold bullion in its years of production, from 1938 to 1972. Marius Madsen, himself, later became President of the company and of other companies doing exploration work in the Red Lake area and on Baffin Island. He was also involved in real estate, hotel, and construction businesses before his death.

Nehemiah Faulkenham, the original staker of most of the land that became the Madsen mine, was one of the most colourful figures in the gold rush. Faulkenham, described by a *Toronto Star* reporter as "a sinewy Nova Scotian who stands about 5'6" in his moccasins",[22] was a Nova Scotian by birth, but by the

----

[22] *Toronto Star*, February 18, 1926.

time of the Red Lake gold rush he had become one of that group of well-seasoned prospectors who floated across Canada from one mining camp to the next. He had been at Porcupine and at Kirkland Lake when those fields were opened up, and, like most prospectors, failed to strike it rich at either time.

Faulkenham first went to Red Lake in January, and it was then that he staked his three claims in the Madsen area. Faulkenham figured in the history of Red Lake for only some eight months, as he was killed by a dynamite blast in August of 1926, but he became a minor legend in that short time: for his stamina; for his generosity, both as a traveller on the trail and then as Red Lake's first storekeeper; and as an apostle of the fresh air creed. Unlike most of his fellow travellers on the trail, to whom personal hygiene was not a major concern, Faulkenham bathed every morning. From the day the lakes opened in the spring until freeze-up in the fall, he would take an early-morning dip; for the rest of the year he would roll naked in the snow and give himself a quick rub-down before dressing and cooking his breakfast.

Another group of mining veterans also mushed in to Red Lake in January, unique in that it was all composed of members of the same family. The McDonough family had been active in exploration and development of mining properties all over northern Ontario. There were six brothers in the family: Jack, Pete, Patrick, Ed, Tim, and Albert. (Jack and Joe, of Madsen fame, are the same brother.) A classic, though fuzzy, photograph taken in February, 1926 shows five of them, plus a friend and a brother-in-law, lined up for the camera at Hudson with their dog team, fur hats on and spectacles gleaming in the sunlight.

Near the end of January, Jack and Pete (in the vanguard of the family compact on this occasion) first went to Red Lake, and staked claims on part of what became the Cochenour-Willans property and on McKenzie Island. Things looked good so they sent out for more brothers to join them, and drove their dog teams out to Hudson to meet them. Jack went on to Toronto and Pete guided the new arrivals in to Red Lake.

When they reached Red Lake, they found everything staked near their own claims, and so carried on to the north-east to Hoyle Bay, then over to Trout Lake and Birch Lake. The party staked close to 150 claims. A few days after they had completed their staking, a couple of the hundreds of new prospectors arriving in Red Lake asked someone where they could find unstaked ground. The newcomers were directed to the Hoyle Bay area where they spent several tiring and disappointing days looking for available land to stake. When they returned, disgruntled, to Red Lake and were asked if they had succeeded, one replied, "Hell, no. There was a McDonough under every pile of brush up there and the name McDonough was on every damned claim post."[23]

The month of January also saw the return of Jack Hammell with a party that included three engineers from Dome (Dome Mines had just become involved in the Howey property) and an Englehart prospector named Charles Henrick. Henrick had happened to meet his friend Doug Wright, Chief Exploration Engineer for Dome, in a railway station and was invited to join the party to Red Lake. (Party was an appropriate term for this expedition in more ways than one, according to Henrick's description of it.) The group had such a happy and well-lubricated journey on the train to Hudson that it took them several days to pull themselves, as well as the necessary provisions, together before they set off on the trail to Red Lake.

It took the Hammell/Henrick group 12 days to make the trip, and food became a major concern. Henrick was in the lead with the best dog team, carrying a varied load which consisted of the Dome's typewriter and office supplies, 100 pounds of salt herring and 100 pounds of corn meal for the dogs, and 20 pounds of pork chops for the men. On their first night out, which they spent in the cabins of the Hudson's Bay post at Lac Seul, a dog sneaked in and ate all the pork chops.

As a result, the party ran out of food, both human and canine, before they reached Red Lake, and supper on the last night on the trail was a slim one all around: a handful of bread

---

[23] The source remains elusive.

crumbs for each man, and a strip of frozen moose hide, which Henrick had found near camp, for each dog. When they arrived at Alex Gillies' tents on Red Lake, there wasn't much food there either, but the enterprising Mrs. Palo was nevertheless able to prepare for them a meal that must have seemed like a banquet: baked lake trout, fresh bread, prunes and tea.

The reason for the arrival of the Dome party was that Jack Hammell had pulled off a financial coup. He had managed to interest the large mining corporation, Dome Mines, in the Howey property, and Dome took out an option on the Red Lake claims. The contract between the two parties was announced by Hammell to the press on January 16. According to this contract, as described later by Dome's general manager, H.P. DePencier, Dome Mines undertook an option to purchase the existing syndicate's 19 claims at Red Lake. To get this option, Dome committed itself to making a minimum payment of $25,000 within 6 weeks of the date the agreement was signed (that date was January 7), a further payment of $50,000, and within two years the sum of $500,000. For these payments, the company would gain a 75% interest in a newly organized company to be called Howey Gold Mines Limited. Hammell's Howey Red Lake Syndicate retained a 25% interest, and the 30,000 units in the new company were to be sold for $5.00 each.

This committment by Dome was the most important factor in the development of a full-scale gold rush from a single, solid but unspectacular find in an inaccessible spot. If Dome thought the Red Lake property was a good one, then good it must be. And the discovery of a potential gold mine was important national news. The mining industry at this time was seen as a major focus of Canadian development, and likely to become even more important to Canada's industrial and commercial life. Indicative of this viewpoint were headlines in the *Toronto Star* of January 5 and 6, proclaiming that "Canadian Mines Will Act as Magnet for Capital, People" and "Steady Advance on Average Price of Mining Stocks in 1925 . . . Gold Shares Lead the Way."

February 1926

The time was ripe. By the beginning of February large numbers of prospectors had begun to hit the Red Lake trail through Hudson. On February 4, the *Manitoba Free Press* proclaimed "Red Lake Mine Rush Biggest in 20 Years" and followed this up with a major story two days later containing the prediction that 1,000 prospectors would be at Red Lake by summer. On February 5, the *Toronto Star* announced, "Great Rush to Red Lake." One of the prospectors who was in Red Lake on the 1st of February announced that about 20 parties were arriving from Hudson every day.

This was mid-winter now. February is one of the coldest months of the year in that area and, in retrospect, it seems nothing short of miraculous that no deaths occurred on the long cold trail from Hudson to Red Lake. During the month of February, several kinds of motor vehicles were tried on the trail to make that long, cold journey easier–and all of them failed.

Conveyances called "snow motors", which initially promised to haul 5-ton loads all the way to the north end of Lac Seul, were tried but had to be withdrawn. The deep slush on the lake gave them no bearing surface, and they could not get a foothold on the portages because there wasn't enough snow. (A similar fate befell the large "snow motors" made especially for the Detroit Arctic Expedition, a group of adventurers who hoped to reach the North Pole via Point Barrow, the northern tip of Alaska, in early 1926. On that expedition, the snow motors took 14 days to go 65 miles and used 400 gallons of gasoline before they eventually broke down.)

Similarly, a Toronto man named Johnny Jones (described by the *Toronto Star* as "Toronto's well-known explorer, adventurer and prospector")[24] designed what he called a "motor sleigh" especially for use on the trail to Red Lake. This contraption, mounted on two pairs of runners and featuring a four-bladed propeller, was supposed to go 50 miles an hour over snow and ice, according to Jones. Unfortunately, the propeller blades were smashed as the motor sleigh was being crated for shipment to

----

[24] *Toronto Star*, February 22, 1926.

Hudson, and the vehicle does not seem to have made it to the Red Lake trail at all. Jones himself, though, reappears in the gold rush story as a "maker of moving pictures" (he was obviously a man of many parts) whose party became lost, four days out of Hudson. They were found a mile off the trail by a police party, "looking for Indians to photograph."[25]

The failure of motorized vehicles to cope with the rigours of the trail left two kinds of power, to which almost everyone resorted at that stage in the gold rush: man-power and dog-power. A large number of would-be prospectors walked in from Hudson, pulling behind them a toboggan with their necessary supplies loaded on it. Among these was Ed Patterson, one of the many on the Red Lake trail who had been to the Klondike in 1898. (The trip to Red Lake obviously had nothing to compare in terms of hardships with the treacherous Skagway route to the Klondike; one of these Klondike veterans remarked that the Red Lake trail was just a "Pullman trip" to him.)

Another man who walked in to Red Lake was Joe Biron, one of the few French-Canadians attracted by the news of gold at Red Lake. This man later distinguished himself in the fledgling community at Red Lake by wearing his yellow curls tied on top of his head with a red ribbon and going naked when in the privacy of the bush. He used to take a daily dip in the lake, summer or winter, and took one such dip, while crossing the ice-covered surface of Red Lake near Pipestone Bay, that was noted with amazement by a passing party. The ice was soft in one spot and Joe's toboggan broke through, sinking rapidly to the bottom. As all Joe's worldly possessions were now at the bottom of the lake, he determined to regain them. He stripped off his clothes on the ice (so that, as he explained later, he could swim "more better"), dived in head first, and came up with the toboggan rope in his hand. He then dressed and proceeded on his way. As someone remarked, "Joe was a hard man to stop."[26]

----

[25] *Toronto Star,* March 15, 1926.

[26] Gordon Shearn, "Early History of Red Lake," *Precambrian,*

One would-be prospector who walked in to Red Lake, and who has left a written record of his trip, is Conrad Hanson, who then spent most of his life at Red Lake.[27] Hanson was a young man who lived in Boston and worked for the Massachusetts State Industrial School, in February of 1926. One morning he opened the *Boston Post* and saw the headline, "Gold Rush at Red Lake". He was smitten immediately.

His reaction to that headline were characteristic of the feelings that stirred in the breasts of the thousands of hopefuls who left whatever they were doing, wherever it was, to go to Red Lake to find gold. "I had been looking for an out, and this had the appearance of being it. Red Lake was in a wilderness almost untouched by man. It was a new frontier in a new part of the world that sparkled with adventure and business opportunities. The next morning I handed in my notice to the personnel office."

When Hanson boarded the Boston train for Toronto, it was raining; when he disembarked from the CN line, 27 hours after leaving Toronto, at Hudson it was 25 degrees below zero. Hanson remarked that he felt sorry for the dogs he saw chained to every post and lying on the snow with no protection from the icy winds. He later discovered, though, that these hardy animals could be perfectly comfortable in conditions much worse than that.

Disembarking from the train with Hanson, as it happened, was a small group that included Bill Jackson and Joe Manion, who became the stakers of the Jackson-Manion Mine at Woman Lake, and Dave McQuaig, later a businessman at Red Lake. After Hanson had breakfast at Hudson, he struck up conversation with as many prospectors as he could, in order to get some information, and from one of them he bought a toboggan which

----

April 1956, p. 12.

[27] This record is found in the book *Gems of the North* by Conrad Hanson, privately published, Kirkland Lake, 1979. I am indebted to this volume for the following story of Hanson's first trip to Red Lake.

he then pulled over to the Hudson's Bay store and loaded with gear which he bought there.

As he was loading it, he was surrounded by a group of seasoned veterans who gave him the benefit of their discouraging words. A number of greenhorns had tried to walk to Red Lake before, pulling toboggans, it seems. They had made it through one day, built a small fire over which they huddled during the night, and–nearly frozen–had turned back to Hudson, sold their outfits, and used the cash to buy a train ticket back home.

(One such innocent, according to the Hudson oldtimers, was a sportily-dressed young man who had stepped off the train carrying a guitar. Thinking, no doubt, of the Klondike saloons, this fellow's plan was to make a few dollars by playing and singing at the stopping places en route to Red Lake, and in the bars at Red Lake when he arrived. The young man was obviously unaware not only of the primitive nature of such stopping places as existed along the way, but of the fact that Prohibition was in force in Ontario at the time. He soon turned back.)

As Hanson finally headed off to the north in mid-afternoon, pulling his toboggan, one of this veteran group remarked to McQuaig, "There'll be another outfit for sale tomorrow." He was wrong; Hanson may have looked like a greenhorn as he stepped off the train in his city clothes, but in fact he was a hardy farm boy by upbringing.

Travel on the first day went well, as the snow surface was hard-packed and the toboggan pulled easily. When darkness fell he cut wood, made a wind-break from the canvas tarpaulin covering his goods and a bed from spruce-boughs, built a large fire, cooked supper and went to sleep rolled in blankets. The second day was much worse, as a fresh fall of snow made the going slow, and because, as much of the area he was passing through had been burned in forest fires, it was difficult to find firewood. He spent two nights in vacant tents that had been erected by crews of men sent in to cut roads across the portages. A night in such a tent was a mixed blessing: the tent did provide shelter and a heater of sorts, but it also provided large numbers of lice who thrived in the woollen underwear and the heavy, infrequently washed blankets. On the seventh day, Hanson reached Red Lake.

Those who were brave enough (or poor enough, or foolhardy enough) to walk in to Red Lake from Hudson were definitely in the minority. The operative word in the gold rush was "dogs". Gordon Shearn, who went in on the trail in March, remarked that some hopefuls en route, by train, to Hudson had painted "Red Lake Or Bust" on the sides of their railway carriages, but this was not the real slogan of the rush. If there was such a slogan, said Shearn, it would have to be "Mush, You S.O.B.'s!"

The first sight, greeting the traveller getting off the train at Hudson, was that of sleigh dogs everywhere. There were dogs of every size, breed and colour tied to every tree in Hudson, from the lake shore to the top of the hill behind the town. There were dogs belonging to the hopefuls who had arrived previously on the train and were getting ready for the trip; dogs belonging to parties that had already been to Red Lake; and innumerable dogs on the train itself. (In fact the CNR, whose service received nothing but praise as it accommodated itself to the sudden boom in traffic, was forced to change its regulations in order to deal with the hordes of dogs. On February 27 it was announced that sleigh dogs that had previously travelled for $3, as excess baggage, would now cost $8.50 crated or $11.00 uncrated.)

Major Cunningham-Dunlop, that valuable source of information, told a *Toronto Mail & Empire* reporter, "So far as transportation is concerned, the gold rush appears to be a case of reverting to the primitive, for all forms of transportation other than that of the dog-team have failed to stand the strain of a trip from Hudson to Red Lake . . . Because of that, the dog has already been quoted at a premium, as the prevailing price for good dogs when [I] left Hudson a few days ago was $200 each. Even so, the demand has been so keen that all the dogs in the district have been brought in."[28]

Huskies, mongrels, breeds of every description were pressed into service. (One observer noted that there was a temporary drop in the canine population, all the way from Winni-

----

[28] Quoted in *Canadian Mining Journal*, February 19, 1926.

peg to Fort William, during the gold rush.) *Star* reporter Paul Reading who mushed in to Red Lake in February, described the gold rush as a dog derby. The dog team which Gordon Shearn contracted in Hudson to drive to Red Lake was a typical one. Said Shearn," . . . [after] our dog team had been duly delivered . . . I am afraid we gazed at them and at one another in silence for a long time. They sure were a nondescript bunch, only one of the six bearing any resemblance to the breed usually associated with dog-harness, and that one so thin you could spit through him. One was definitely an Airedale and as for the rest, anybody's guess was good."[29]

Regardless of size or breed, then, the dogs (usually in teams of six) were harnessed, hitched up to heavily-laden toboggans and driven off. In every case, much of the load they were carrying was their own food. The dog teams, ill-assorted as they often were, could perform with enormous strength and stamina and needed no coddling–but they did need food, and a lot of it. The standard daily fare on the trail for a working dog was 2 or 3 pounds of boiled cornmeal plus a few ounces of fat, or else 3 or 4 pounds of frozen fish.

Supplies of these commodities were easy enough to obtain at Hudson before going in, but naturally dwindled quickly at Red Lake. For a time dog-food there was very scarce. Not only did the trading posts run out of cornmeal, but the fish in the lake seemed to have deserted their usual haunts. Major Cunningham-Dunlop told the *Toronto Star* that he had to feed his dogs "biscuits" on the way back to Hudson in February, and that it cost him $5.00 a meal to do so.

Travel by dog-team, in 1926, was a common feature only in the far north of Canada, and the fact that it was the major mode of conveyance to Red Lake gave this gold rush much of its romance. On the other hand, "dog derbies" were still a major sporting event. The winning team from one such derby at Le Pas, Manitoba, in January, took a highly-publicized train trip to Quebec, in February, to compete in an international derby there.

----

[29] Gordon Shearn, "Early History of Red Lake," *Precambrian*, April 1956, p. 10.

Dogs, were, however, naturally endowed with coats to ward off the extreme cold, whereas their drivers were not. The standard set of clothing for the trip consisted of heavy woollen underwear, a mackinaw suit (shirt and pants in a blanket-like wool fabric), a pair of workmen's overalls, a parka with hood that would pull out to protect the face, 6 pairs of woollen socks and moccasins. There was little variety in this basic outfit, though Neh Faulkenham appears in one photograph with a rather surprising striped parka over all. Most of these layers of clothing would be worn all night as well, for many of the travellers slept in the open while on the trail, or else in unheated tents.

Most travellers spent their first night out from Hudson at the Lac Seul Hudson's Bay Post. Here, for 25 cents a night, they could sleep on the floor of what was known as the Indian House, a log building previously used only by visiting trappers and empty except for a wood stove. Party after party of men would stumble in, half-frozen from the trail, to hang their wet mitts and socks to dry over the stove, unpack their bedding, and, by the light of coal-oil lanterns, unpack their simple food and line up to cook it at the stove.

The second night was usually spent further up Lac Seul at Sand's Fish Camp, a group of buildings that had been rented by an enterprising fellow from Sioux Lookout, for the winter months. This entrepreneur made a tidy profit by charging a dollar a night for sleeping space (on a pole bunk if you were lucky, on the floor for the same price if you weren't), and also providing firewood and meals, for a price.

Further on, a prospector might be able to avoid sleeping in the open if he were lucky enough to hit a teamster's tent, like the one Conrad Hanson slept in, or one of various dilapidated cabins to be found in the woods, which provided some shelter, like those of an old fur-farm just south of Red Lake. The standard food on the trail was no more luxurious than the accommodation: beans, bacon, bread or bannock, the lot washed down by strong tea.

As the new arrivals streamed in to Red Lake in February, all word from the Howey was positive. Dome Mines optioned another 20 claims, these ones from the Cochenour and Cockeram syndicates. This further evidence that the remote Red Lake area was of interest to a large corporation like Dome, made the prospects there even more attractive. Lorne Howey arrived in Hudson from Red Lake on February 16 and reported on the work that had been accomplished on the property to that point. Twenty-three trenches had been dug, some of them 60′ in length, and 930′ of gold-bearing vein had been uncovered. After dynamite blasting had been carried out in the trenches to explore the surrounding rock, free gold was found in every trench.

There were no extravagant showings of gold from this preliminary exploration, but gold was disseminated throughout the vein and the sample assayed at $9.00, or .44 ounces, per ton. This could be considered a moderately rich ore. (The actual ore processed by the Howey mine was much less rich than that, however, decreasing in value from .20 ounces per ton in 1930 to .06 ounces per ton in 1940, and the mine made profits at those levels. By comparison, the ore milled at Campbell Red Lake, one of the highest-quality ores milled in Canada, averaged an outstanding .59 ounces per ton in 1981.)

The *Sunday Times* of London, England on February 7 carried an article entitled GOLD RUSH: LONG TREK OVER WINTER TRAILS IN CANADA, and newspapers throughout Canada began to carry frequent references to the Red Lake rush. The time had come for this sudden, primitive migration to Red Lake to be recognized not just by the media but by the various levels of government, and the month of February saw the first establishment of government institutions at Red Lake. In this month the tent settlement at Red Lake received attention from not one but three representative bodies of His Majesty's government: the Ontario Department of Mines, the Ontario Provincial Police, and the Canadian postal service.

For several months, the Ontario Department of Mines and its Minister, Charles McCrea, had been under pressure to establish a Mining Recorder's Office at Red Lake. Under the existing system, men who had taken the long, cold trek to Red Lake to stake claims then had to go all the way back out to the recording office at Kenora to record those claims.

The Department, in fact, reacted fairly quickly to the new developments at Red Lake, as the *Toronto Star* announced as early as January 21 that, "in order to assist the prospectors and others working in the Red Lake area, the department of mines has decided to establish a branch recording office in Red Lake, open for business the early part of February." That proposed opening date may have been just the reporter's speculation, for it would have been impossible to assemble and transport the necessary men and supplies in so short a time.

Two officials were to go to open up the new branch: these were Thomas A. McArthur from North Bay, who was the Superintendent of Mining Recording Offices in Ontario, and Herbert Holland, the Mining Recorder at Kenora. They met at Kenora to gather together the essential supplies and equipment; transported their load 18 miles north to Redditt Station on the CN line; and then went east by train to Hudson.

February 19 was the day the Mining Recorder's party departed from Hudson, and, with their party, travelled the first representative of law and order to take up residence at Red Lake, Constable Lawrence Carr of the OPP. With respect to law and order (as in all other respects), the gold rush to Red Lake was a very typically Canadian experience. As in the settling of most of the other frontier areas of Canada–usually in the previous century–the presence of the law was already established at Red Lake when most of the people arrived. And again, typical of the Canadian frontier experience, the job of the law's representative was straightforward. The mining camp that existed at Red Lake in 1926 was not distinguished by any outbreaks of violence or even by much rowdiness; Carr was reported by the *Toronto Star* on March 4 to have had "no trouble maintaining law and order in [this] latest outpost". Holland, the Mining Recorder, became the mining camp's first magistrate as well, and the most common (and most heinous) offence by far at Red Lake was bootlegging.

The Mining Recorder/OPP party was joined by a *Toronto Star* reporter, Paul Reading, who coincidentally arrived in Hudson (from the opposite direction) at the same time as the other group. The Holland group had pre-arranged transportation to Red Lake with the McIntyre Mines' regular driver, who had lined up another team and driver as well, but when this fellow saw all the equipment and supplies the official party had with them, he realized that two teams were not enough. Just at that time Neh Faulkenham happened to arrive at Hudson en route from Red Lake to Kenora to record his claims.

Faulkenham had one of the best dog teams in the area (this was part, but not all, of the reason why he is reputed to have covered the trail to Red Lake faster than anyone else, in 4 days). He had also probably been over the trail more often than anyone else, and the fortunate party set off for Red Lake with Faulkenham as guide and leader of the group, in the morning of February 19.

Because reporter Paul Reading was one of the group, their trip on the trail to Red Lake was much better documented than most. Reading's first reports of the journey were published in the *Toronto Star* on March 1st. (And many of them were picked up by other papers, including the *Manitoba Free Press.)* They were front-page news, accompanied by a map showing possible routes to Red Lake, and headlined "PARKA-HOODED CARAVAN MUSHES ON ICY GOLD TRAIL; STAR MAN ONE OF PARTY". His "experiences and observations on the Arctic trail" include his comment that the gold rush to Red Lake "will be remembered not for the swiftness of the race but for the intensity of its effort. . . . There was no pause to exchange the courtesies of the road. . . . There are claims to be staked ahead, and first choice falls to the first arrivals . . . " (Ironically, of course, all the good claims had been staked several weeks before this anyway.) The long surface of Lac Seul was slushy, apparently, and Reading commented that "none of us came through dry-shod, and Constable Carr had two of his toes frostbitten."

Reading's description of the settlement at Red Lake is the best contemporary one there is. Dated February 24, it reads, "Tucked away in one corner of a bay near the outlet of this straggling sheet of frozen water is a cluster of tents. They are pitched on hastily constructed log cribs and roughly floored with hewn spruce. Through the roof of each a section of rusty stovepipe thrusts. . . . One of the tents bears above the flaps the end of a packing case bearing in staggering black letters the . . . legend 'mining recorder's office' [it didn't open for business until March 1]; another, though it does not yet boast a sign, is a provincial police post, a third is the 'office' and a fourth the cook tent . . . " "To the little poplar trees that stud the shore, a legion of dogs are chained. . . . They fill the air with doleful sound . . . " "But through this chorus of mourning rings another sound . . . the unmistakeable tapping of steel upon rock . . . it is the noise of the samplers chiselling channels across the underband of quartz where Lorne Howey struck gold. . . . The chips as they fly are . . . collected, bagged . . . and labelled for the assayer . . . " "This desolated community assures itself that it is to form the nucleus of a great camp. . . . This will be a stampede probably unequalled since the Klondike."

(The glowing optimism of Reading's predictions was the norm for media coverage at this time, with the exception of one paragraph buried in an article on February 15 in the *Toronto Star*, that seems to have deterred no-one from rushing up to Red Lake. This unusual note of warning commented that "It is blind chance that guides [the stakers] at present, for a guess is all you can make at the treasure that may rest beneath a 4' blanket of snow. Those who have been pouring in since last December have little chance of knowing what they have staked. Most of them follow the formation of the rock as government geologists have charted it; get in as close to the original discovery as they may; and let it go at that.")

It was while the Holland party was en route to Red Lake to set up the Mining Recorder's Office, in fact, that the Ontario government announced that Red Lake would not only have an office, but would become a separate mining district as of March 1st. And, according to the press, the bill to create the Red Lake mining divisiion established a record for speed in its passage through the legislature. (This fact was quaintly headlined in the *Toronto Star* of February 26 with the words, "LEGISLATORS JOIN IN 'RUSH' TO AID MUSHERS OF RED LAKE.")

The first appearance in Red Lake of a representative from His Majesty's Post Office is not so fully documented, though it is known to have occurred in February as well. Bill Brown, who was postmaster at Allan Water, Ontario (a village on the CN line about 80 miles east of Sioux Lookout) was sent in to Red Lake to investigate the postal requirements of this rapidly growing settlement.

Bill Brown was an interesting character with an unusual background. As a young Englishman he had served in the Great War, like many others who mushed in to Red Lake, but he is the only one known to have been a student at Oxford prior to his war service. (He did not, however, graduate, but was apparently sent down for some misdemeanour and sent by his family to work in Sweden.) After the war he joined the service of the Hudson's Bay Company and came to Canada, where he quickly rose to the position of Factor at Osnaburgh House on Lake St. Joseph.

His tenure at Osnaburgh was short-lived, though, and he is reported to have the unhappy distinction of being the only Hudson's Bay factor to be fired direct from London, England. Apparently one day Bill left a lighted cigarette lying on a sloping counter in his store; the cigarette rolled off and fell into a nearby keg of gunpowder, and the resulting explosion demolished the store and killed two of the Indian clerks.

Bill was badly wounded in the explosion, but was pulled out of the debris by an Indian girl who nursed him back to health, and whom he then married. He was relieved of his duties to the Hudson's Bay Company, and went to Allan Water where he worked at a sawmill and then at the post office. His report on the necessity for postal service at Red Lake obviously came out in favour of it, as Red Lake became one of the first communities in Canada to have airmail service, when regular flights to the tent community on Howey Bay were established at the beginning of March. (Before that, there was no postal service at all to Red Lake: resident trappers, traders, and prospectors had to go out to Hudson to pick up their mail.)

During the month of February, plans were afoot for the air service that was soon to begin, but there was another non-motorized means of locomotion that was used on the Red Lake trail, in addition to the more common dog-power and man-power. This was horse-power. Several teams of horses and drivers had been hired by Dome Mines to go to the north end of Lac Seul, and to construct rough log roads across the portages.

There were other teams for hire at Hudson as well. After Conrad Hanson reached Red Lake on foot, he found his train companions already in residence there. They had made the trip so quickly because they had hired a team of horses and a sleigh to transport all their supplies as far as Pine Ridge, then hired Indian drivers and dog-teams to bring their supplies the rest of the way.

Another prospector who came over the gold rush trail by horse was Tim Crowley. Tim, whose home was in Quibell, was a veteran of bush and mining life in northern Ontario who had worked on the railway, in the mines and at prospecting all his life. (He was also a veteran of the Great War, having persuaded the officers concerned that he was under the age of 45 when he enlisted in 1916, although he was actually 62 at the time.)

Several times in his long career, Tim had come close to making the big find every prospector dreams of–but never quite close enough. He first came to the Red Lake area in 1912, when he and his three sons discovered and staked the iron ore body on Bruce Lake, to the east of Pakwash Lake, south of Red Lake. (The claims lapsed and changed hands several times, and this property was eventually brought to production 44 years later by Stelco, which operated a large iron ore pelletizing plant there.) Crowley came back to Red Lake in the minor flurry of activity there in 1922, when he staked the claims that, after lapsing and being re-staked, eventually became the Gold Eagle mine which produced about $1.5 million in gold between 1937 and 1941.

When the real Red Lake rush began in 1926, Crowley determined to go in again. At this time, though, he may have been feeling his advancing years (he was 72 at the time), for instead of going by foot or by dog-team, he hitched a horse to his fifteen-foot toboggan and set off north from Quibell to the English River, about half of the distance to Red Lake. (Quibell is about 75 miles west of Hudson.) It took him fourteen days to get as far as the English River, as the snow was deep and the trail unbroken, and by the time he reached that point he was running short of supplies. He returned to Quibell instead. There he heard that everyone was going in from Hudson, so he took his horse and toboggan there and headed north, reaching Red Lake at the end of February.

Crowley's horse, and all the other horses on the trail, were daily sources of irritation to the hundreds who were mushing in by dog-team, as the horses' hooves cut up the smooth surface and made the going very rough for the dogs. Some of the drivers of horses, however, had their own causes of irritation. One such would-be prospector, Jess Pond of Sioux Lookout, set out from Hudson with a horse pulling his load. On the first day out, unfortunately, the horse broke away and continued on the trail on his own. Pond announced that he was still going to Red Lake–and that if he found the horse there when he arrived, he would turn it into dogfood.

All the hopefuls who went in to Red Lake in February went by primitive conveyances, but there were grand plans for better means afoot at the time. (Of these all, only the plans for air service materialized.) The *Manitoba Free Press* reported on February 6 that there were rumours of a road to be constructed from Jones Siding (on the CN line, about 15 miles west of Quibell) to the English River. There were also rumours of a wagon road to be constructed from McIntosh, on the CN line almost directly south of Red Lake. The *Free Press* also reported that a Kenora-Red Lake railway was in the works. (All these routes, like that of the T & NO Railway in northeastern Ontario, were intended to promote agricultural and timber development as well as mining.)

The Manitoba newspapers, naturally enough, were concerned with transportation connections to Red Lake that would eventually flow through the city of Winnipeg: either south from Red Lake and then west to Winnipeg along the railway lines, or else west from Red Lake to the mining areas of central Manitoba and then south to Winnipeg. If these proper connections were made, it was felt that the development of the Red Lake mining area would mean as much to Winnipeg as the development of Cobalt/Porcupine had to Toronto. To Winnipeg, Red

Lake appeared to be part of what the Winnipeg press referred to as the "mid-continent mining field", separated only by the accident of a provincial border from the mining areas just to the west of it, in central and northern Manitoba. To Toronto interests, on the other hand, Red Lake obviously seemed to be another Porcupine, just a few hundred miles to the west of the first, and, therefore, Red Lake's obvious links would be to Toronto.

The *Toronto Star* on February 18 reported that Lorne Howey had returned from an interview with Charles McRea, the Minister of Mines, feeling "quite elated" because "they are finally deciding to push a road into the Red Lake district from Jones, Quibell, Hunter or McIntosh, and the work will soon be under way."

In the event, there was cause for neither concern nor elation. There was no road built into Red Lake until the end of the 1940s. The only railway line ever built north of the CN line in that area was a 68-mile spur line built to serve, not the gold mines, but Stelco's iron ore property at Bruce Lake.

During the month of February the newspapers were filled with current plans for another kind of transportation service to Red Lake, by air. These plans succeeded, where others for road and rail service did not, principally because the capital requirements were much less and because a functioning air service could be put into operation by relatively few people.

There were two separate sets of plans going for air service. The first was the brainchild of Jack Elliott and Harold Farrington. Farrington had served with the Canadian Army in World War I and, as he told it, was in the trenches in France one day when he saw a plane flying overhead and decided that the air was the safer place to be. He transferred to the R.A.F., went to England for flying lessons, and flew as a second lieutenant for the duration of the war. His first "aviator's certificate" was issued in June, 1918 by the Royal Aero Club of the United Kingdom, and he has spent the rest of his life in aviation–and most of it at Red Lake.

After his work with the Elliott service at the time of the gold rush, Farrington flew for Western Canadian Airways; this included flying mail from Hudson to Red Lake, flying an experimental daily air service between Winnipeg and Calgary, and making several historic flights to the Arctic. He made innumerable mercy flights during his career. He did bush flying for various operators both before and after World War II, when he flew with the RCAF. He ended his long and honourable career in Canadian aviation as an airline's agent in Red Lake, and still lives there–a wiry little man who still walks several miles daily, who has a remarkably precise memory, and who misses flying, he says wryly, just the same as he misses everything else.

In February of 1926, Farrington was working in Hamilton for Jack Elliott, who was operating a flying school from the ice of Hamilton Bay. Business was slow, and when Farrington read of the gold rush to Red Lake, in a local newspaper, and saw the primitive modes of transportation being used by the people rushing to get there, he thought that air service to Red Lake might be a profitable business to start.

He discussed it with Jack Elliott, who was in agreement. Elliott had purchased ten aircraft for his flying school. These were open-cockpitted craft, Curtiss JN4 "Canucks" popularly known as "Jennys", which had been built during the First World War. Each Jenny could carry, in addition to the pilot, a cargo of 200 pounds, and the Jenny's air speed was 90 miles per hour. Elliott decided to inaugurate an air service to Red Lake with three of these aircraft.

It seemed wiser to ship the aircraft by train for as much as possible of the distance between Hamilton and Red Lake, as regulations at the time required that an aircraft be given a complete overhaul after each 100 hours of flying time. The aircraft, accordingly, were padded, crated, loaded into boxcars, and shipped from Hamilton, on February 17. Elliott and Farrington had originally planned to use the closest point to Red Lake as their air-base (McIntosh and Quibell vied for this honour, one being 73 and the other 76 air-miles from Red Lake), but after checking out those two desolate whistle-stops, they opted for Hudson instead. Since all the prospectors were going in from there, the route to Red Lake was well-marked–and well-peopled as well–in case the planes should encounter any trouble and be forced to land.

It was February 26 when the aircraft arrived at Hudson, and they were immediately uncrated, assembled, and put through test flights by an Inspector of the Transport Civil Aviation Branch. (Three Jennys were shipped from Hamilton, but only two were flown on the route to Red Lake. The third, for whatever reason, seems to have been used for parts.) Captain Farrington, who was to pilot the aircraft, suggested that skis would be a good addition to the tail assembly of the planes, which often dragged in deep snow. He had two small skis made for this purpose by a firm in Sioux Lookout; they were bolted to the planes' tail assembly as directed, and proved satisfactory. This spur-of-the-moment innovation later became standard procedure.

Elliott and Farrington plotted their air route to Red Lake, generally following the mushers' trail most of the way to Pine Ridge, at the end of Lac Seul, but then taking a more direct route to Red Lake. They worked out the distance to be an even hundred miles. Their fare was set at a dollar a pound, one way, passengers or freight, and they had only to wait for clear weather to begin flying.

It was in the latter weeks of February, as well, that plans were being made for another air service to Red Lake. This business was being planned jointly by Captain Harry Oaks (not to be confused with the mining magnate of the same name) and Major G. A. Thompson. Oaks, who was a mining engineer as well as a veteran of the Royal Air Force, had flown the Curtiss flying boats into Red Lake the previous fall under contract with Jack Hammell; Major Thompson had also flown both in World War I and for the Ontario Provincial Air Service. These two had, along with a third partner, gone into Red Lake by dog team shortly after Christmas, staked some claims, and sold them in Toronto for enough money to make a down payment on a Lark aircraft.

The Lark was similar to the Jenny, but could carry 400 pounds in addition to the pilot, rather than the Jenny's 200 pounds. (Or, two passengers rather than one.) The Lark which they bought for their fledgling company, Patricia Airways Exploration Limited, was in Newark, New Jersey at the time. Getting it from there to Hudson proved to be such a problem in the late-winter weather that their air service didn't begin until April 12.

All these plans for air service were seized upon by the press as hot news, as they combined two topics of great interest at the time: aviation, and the gold rush. Flight was still in its early stages. In the whole of Canada, there were only 44 licensed aircraft in 1926. The first regularly scheduled air service in Canada had begun only in 1924 (and that too was in an isolated area, the Rouyn mining district). The newspapers at the time carried almost daily stories of flights being made, or attempted, all over the world.

In the spring of 1926 Captain Amundsen and Lincoln Ellsworth attempted to "fly across the top of the world", from Norway to Alaska via Spitzberger, by dirigible. In a separate expedition, Commander Byrd of the U.S. Navy went to Spitzberger via ship and thence attempted a flight across the Pole, through Greenland. When the Detroit Arctic Expedition flew over the Alaska mountains, newspapers contained a blow-by-blow description of the flight filed by an intrepid reporter sitting in the cockpit. Another newsworthy event occurred when an American, John Goldstrom, set off from New York by ship to try to break the previous record for circumnavigating the globe (35 days, 21 hours, 35 minutes) in an epic voyage that combined air and sea travel.

Anything concerning flight was news–for example, WOMAN OF 104 FLIES IN AIRPLANE was a headline in the *Star* of April 27. The speed of airplanes was still being tested at this point: in what was described as a "novel contest" in Auburn, New York, an airplane proved to be 23 1/2 miles per hour faster than the fastest pigeon in a race between two planes and 59 homing pigeons.

The combination of flight and gold rush was a winning one as far as the media was concerned. The *Toronto Star* on February 26, for example, carried a long article, complete with photographs, headlined "BUY TROPHY-WINNER PLANE FOR RED LAKE AIR SERVICE: Major Thompson and Captain Oaks Purchase Wright-Bellanca Monoplane, Winner at New York Races,

To Inaugurate Air Service Into Mining District March 10." And on the following day it declared, "CAN FLY TO RED LAKE WITHIN 2 HOURS: Airplane Service Into New Field Starts Monday, March 1st," and went on to report that "Jack Elliott has shipped 3 planes from Hamilton to Hudson and 2 will make their first trip on Monday." The *Manitoba Free Press* also carried reports of the imminent air service, but, unlike other newspapers, reported that "people are skeptical about [the planes'] success", as a forced landing on any of the lakes would be difficult because of the slushy snow.

Jack Elliott's aircraft, poised to go at Hudson, also were the subject of the first coverage of the gold rush in two prestigious American newspapers, the *New York Times* and the weekly *Christian Science Monitor* of Boston. (It had been covered previously in many other American newspapers.) The *Christian Science Monitor* ran an article datelined Toronto, February 25, under "GOLD FIELD TO HAVE AIRPLANE SERVICE" saying that "An air service which will carry passengers, freight and later mail is being organized from Hudson, Ont. to the location of Ontario's latest gold strike at Red Lake. The new gold field has caused a stampede that has not been equalled in Canada since Porcupine."

And the *New York Times,* proclaiming "PLANES TO CARRY GOLD SEEKERS TO RED LAKE STRIKE IN ONTARIO", began its article in a lighter vein by saying, "It's going to cost fat prospectors plenty to travel to the new gold strike region at Red Lake when an airplane passenger service is inaugurated out of Hudson on March 1. 'We will charge so much per pound,' announced Jack Elliott. . . . The airplanes . . . are expected to fly the distance in two hours, compared with eight days by dog team."

Everything was ready except for the weather. Lines of would-be passengers for the new air service were beginning to form, with some prospectors so anxious to be on the first flight that they offered Jack Elliott $200 extra for the privilege. On March 3, the skies finally cleared and the first flights triumphantly took off from Hudson.

The Gold Rush, Part III: March, 1926

It was in the month of March, 1926 that the gold rush to Red Lake reached its peak. Everything at Red Lake was looking very promising; the area seemed to confirm the hopes that northern Ontario was a treasure chest of enormous wealth, and Premier Ferguson remarked that, although he might not live to see it, there would not be many crops of daisies over his grave before old Ontario would find she was overshadowed by the north.[30]

It was in the month of March that commercial air service to Red Lake began; that the Mining Recorder's office opened and the staking and recording of claims reached its high point; and that–as far as can be determined–the largest number of prospectors arrived. It was also in March that national and international press coverage of the gold rush became a daily event, and that the first successful transport of heavy freight from the railway to Red Lake was begun.

The clear weather that Jack Elliott was waiting for finally arrived on March 3. On that day the two planes of the Elliott air service flew from Hudson to Red Lake together. Elliott piloted one, and carried as passenger J. C. Rogers, a mining engineer from Toronto, working for the Huronian Belt Mines. Harold Farrington piloted the other Jenny, carrying J. Adair, promoter for the same company as his fellow passenger. In the open cockpit of the Jenny, the passenger sat behind the pilot, and both wore goggles (in addition to a great deal of warm clothing, as the flight to Red Lake in winter weather was a very chilling experience).

Harold Farrington has since remarked that he doesn't know which plane landed first on the deep snow of what is now Howey Bay, and that they weren't expected at that time. But their arrival was cheered by the prospectors who happened to be there to greet them. Jack Hammell appeared shortly, and asked Elliott if he would fly he and a financial associate, Charles Hamilton, back to Hudson. Elliott agreed to do so, but was not

----

[30] *Manitoba Free Press,* March 29, 1926.

sure that the planes would be able to take off in the deep snow. The problem was solved by O.P.P. Constable Carr, who suggested to the small crowd gathered around that they tramp down a runway for the plane to use.

When the improvised runway was ready, Hammell and Hamilton boarded the two planes, each carrying a twenty-pound lake trout destined to feed, in style, their associates in Haileybury and Toronto. When Hammell arrived at Hudson, he sent a wire to the Ontario Minister of Mines, telling him that the first commercial flight to Red Lake had been successfully completed.

The press was only one short step behind that historic flight to Red Lake on March 3; the passenger on the following day's flight to Red Lake was William Scott, a reporter for the *Toronto Star*. (In fact, Scott announced in his front-page story the following day that he was "the first passenger on the Jack V. Elliott air service", which was stretching the truth by one day.) As a result of Scott's presence on the flight, it was recorded in minute detail.

This flight did not begin as auspiciously as the previous day's flights, however, as the engine of the Jenny they planned to fly would not start. On the previous night the temperature had dropped to -20 degrees fahrenheit. In spite of the fact that the engine had been run at intervals throughout the night, it would not turn over in the morning. In his lengthy article in the *Star*, reporter Scott detailed the various measures undertaken between 6 a.m. and 1 p.m. to get the engine going.

The plane was turned so that the morning sun would shine on the engine; boiling water was poured into the radiator; the spark plugs were taken out and heated; the anti-freeze solution and oil were warmed before pouring them in. After about five hours of such manoeuvres, the engine finally turned over. (This episode resulted, in fact, in Farrington's and Elliott's construction of a device which became widely used in northern flying. The two men improvised a portable nose hangar of wood and canvas in which a stove could be kept burning overnight to prevent the plane's engine from freezing.)

The plane, with Harold Farrington as pilot and Scott as the passenger, finally left Hudson about 2 o'clock. They carried a light cargo of snowshoes and sandwiches, in case of a forced landing. Their stay at Red Lake was a very brief one, and when it was time to leave, there was once more some concern about takeoff–this time because the plane was caught in slush. However, they managed to get the plane out and to take off successfully. As a result of the late start they had made, the sun had set before they returned to Hudson, with two consequences: one, that Scott received a frost-bitten knee because of the intense cold that set in after the sun's heat was gone; and the other, that the runway at Hudson had to be illuminated by torches for their landing.

On the next day, Friday the 5th of March, another representative of the media flew in to Red Lake. This was Frank Holmes, a Winnipeg-based photographer who was a stringer for Fox News. Holmes went to Red Lake to shoot a newsreel of this gold rush that had provoked such great interest around the world. Holmes was the passenger in the plane piloted by Harold Farrington, while Howard Watt, another pilot in the new air service, flew the other plane. Just as they reached Red Lake, around noon, a heavy snowstorm developed and visibility dropped drastically. In the course of landing, Farrington's plane ran up on the shore of Howey Bay before it stopped, Watt's did the same on the shore of a small island, and both planes were slightly damaged.

Reporter William Scott, back at Hudson, after hearing of the incident, described it to *Star* readers on March 9 in dramatic terms. "SWIRLING SNOW INFERNO ENVELOPED TWO PLANES ON FLIGHT INTO RED LAKE", ran the headline, and the article went on to say that after the forced landings, "As a movie photographer, Mr. Holmes' first thought was to get pictures, so he jumped out of the plane, landed in a deep snowbank up to his waist, and filmed the machines."

Farrington, Watt and Holmes stayed in the Dome camp while Farrington's plane was being repaired; this involved sewing patches over the torn spots on the wings and filing off the rough edges of the propeller, and the repairs were not completed until noon on Sunday. Farrington then flew Holmes back to Hudson, where Holmes caught the train to Winnipeg and had his motion picture film developed. It was shown for the first time on the evening of Monday, March 8 in Winnipeg, billed as "SCOOP: FIRST MOTION PICTURES FROM RED LAKE."

This was the only motion-picture film shot at Red Lake during the gold rush. The Holmes film is, unfortunately, lost in a large quantity of uncatalogued material at the National Film Archives of Canada. Harold Farrington saw the newsreel shortly after it was made, and recently remarked that it was interesting to see himself landing his plane on the ice of Howey Bay. Other motion-picture film was shot of the Red Lake gold rush, but at Hudson: an Ontario government photographer shot film which is still in viewable condition, and an American cameraman shot film for Pathe news which was subsequently (like the Fox newsreel) shown in the United States.

While the planes were held up at Red Lake being repaired, the world outside, particularly the hungry press, didn't know what was causing the delay. The *Toronto Star's* headline on March 8 proclaimed, during this interval, "PLANES STILL LOCKED IN WINTER'S ICY GRIP–UNABLE TO LEAVE RED LAKE."

The first week of March also brought the first air-mail service to Red Lake. The date of that first airmail flight to Red Lake–that is, the first time a letter was carried by air, with official Post Office approval, and for a set fee–is usually given as March 6. This can be so only if Farrington's plane, damaged from its forced landing in the Friday snowstorm, flew back to Hudson on Saturday, rather than Sunday as was reported.

In any case, when his plane flew back from Red Lake to Hudson, it carried a letter affixed with a special sticker reading "Jack V. Elliott Air Service" and countersigned by the pilot. (All letters flown on the air service were so countersigned, and the required airmail stickers cost 25 cents.) The letter in question was an urgent request from Mining Recorder Holland to the Department of Mines in Toronto for more pre-printed claim recording forms. After being flown to Hudson, the letter travelled by CN trains and reached its destination on March 10. Whether the letter's journey, in fact, took 3 days or 4, it compares favourably with the time required for a letter to travel from Red Lake to Toronto by today's all-air mail routes.

On Farrington's next flight, but one, to Red Lake, he carried a new propeller for Howard Watt's plane, so that the air service again had two planes flying the Hudson-Red Lake route. He also, apparently, carried a bag of oats that he had been asked to bring in to feed someone's horse–and thereby hangs a tale.

This story, which appears in virtually every set of recollections about the Red Lake gold rush ever committed to pen and paper, might be entitled The One Hundred Dollar Bag of Oats. Apparently Tim Crowley (known as "Old Tim", for understandable reasons, as there weren't many men of 72 who mushed in during the rush) had run out of oats for his horse, and asked the Elliott Airways pilot to bring him in a bag of oats on his next flight. The pilot did so, and presented Crowley with not only the oats but a bill for $100. In response to the look on the prospector's face, the pilot explained that flying was an expensive proposition. When the pilot was turning his plane around for takeoff, it became stuck in the slush and the pilot had to enlist the help of Crowley and his horse to pull the plane out. When the pilot asked Crowley how much he owed him, Crowley replied, without batting an eye, "One hundred bucks. Feeding a horse up here's a damn' expensive proposition."

This story has been disputed, however, and it has been maintained that the horse came before the oats–that is, that first Farrington's plane was pulled out by Crowley's horse and then Farrington flew in the bag of oats. In this version, no money changed hands, and both moves were made out of kindness alone. A more worthy story, perhaps, but less interesting.

In any case, the two planes began flying regularly to Red Lake in the week of March 8, and the service lasted until spring breakup late in April. There were usually lines of 10 or 12 prospectors at Hudson waiting for the next flight, and eager to pay the substantial fee of a dollar a pound. (Many of them, in fact, wanted to jump the queue by paying more, but weren't allowed to do so.) Fares were reduced on March 15 to about half their previous level; the new fare was a flat $100 per person, plus 50 cents a pound for baggage. This reduction occurred when the Elliott Air Service was amalgamated with the Fairchild Air Service of Montreal to become the Elliott-Fairchild Air Service.

(Presumably Elliott received an infusion of capital in this manoeuvre which allowed him to drop his prices.)

During the few weeks when the fare into Red Lake was determined by the passenger's weight, a number of stories originated that became part of the standard coin of the gold rush. In the anecdote's basic form, a gold-seeker would hop off the train at Hudson, join the line-up for the air service, get weighed in with his baggage, and pay the dollar-per-pound fare. He would then amble over to the Hudson's Bay Store to buy the necessary heavy clothing for the outdoor life in Red Lake, and put on his new garments. When he came back to board the plane, the eagle-eyed air service representative would insist that the prospector get weighed again, and the hapless chap would find himself paying a substantial sum again.

A variation on this story concerns Jack Munro, an ex-prize fighter, turned mining promoter, who was a well-known and highly respected figure at the time. Munro had a crew of men staking claims at Gullrock Lake and was flying in to direct the operation. But even he was not exempt from the embarrassment of re-weighing, as he found when, after weighing and paying, he slipped over to the Hudson's Bay store and returned with a short piece of angle iron (from which iron plates were made to fix to boots to prevent slipping on ice). The angle iron cost him an extra dollar before he could board the plane.

On March 15, Jack Elliott announced in local newspapers that since the airplanes were flying over the mushers' trail daily, any prospector who was in distress should lay out a signal of green spruce boughs on the ice and a plane would stop for him. The first emergency to require pilot-prospector co-operation, however, had the men on land doing the helping. On the morning of March 27, pilot Howard Watt took off from Hudson with his passenger Jack Hill. Shortly after takeoff an up-current caught the plane; it lifted, descended too steeply, and crashed into the ice surface of Lac Seul. Apparently the undercarriage buckled and both wings were sheared off. Hill suffered a broken arm, various cuts and bruises and a concussion; Watt was also injured, but less severely. (However, the ever-ready William Scott of the *Toronto Star*, staying at Hudson, said in his article entitled "CRASHED IN CURRENT ON LONE GOLD REACH OF LOST LAKE SHORE", that Watt was "unconscious for 19 hours.")[31]

Conscious or not, Watt and Hill were fortunate enough to be discovered in their plane shortly after the crash by a passing party of prospectors who loaded them onto their toboggans, set their dog team in motion and carried them back to Hudson. There the two men were treated in the Red Cross hospital car. Since more serious treatment was required for Hill, the nurses at Hudson telegraphed to Sioux Lookout, from whence a doctor, in a CNR switching engine and caboose, came to take Hill back to Sioux Lookout hospital.

In the course of his interview with reporter Scott at Hudson, Howard Watt remarked that, in his opinion, it was really cheaper to fly in to Red Lake than it was to go by dog team. He explained that it cost $100 (and 50 cents per pound for baggage) to fly in, and the journey took 2 hours at most, whereas it cost about $500 and took 6 days to drive in a dog team to Red Lake. (The $500 figure is on the high side, but it could be reached easily enough if the gold-seeker paid the going rate for dogs at Hudson.)

While Jack Elliott's Jennys (frequently referred to by prospectors as the "old crates") were making their daily hops from Hudson to Red Lake, the lone aircraft of the Red Lake area's second projected air service was attempting to reach Hudson from the opposite direction. The plane in question was a Curtiss Lark that Harry Oaks had bought and the pilot was Roy Maxwell. But, getting the plane from New Jersey presented much greater problems than its owners had anticipated. (The difficulties of the journey, in fact, made Elliott's shipping of his Jennys by train from Hamilton to Hudson seem a masterpiece of strategy.)

The Lark's voyage received very detailed coverage by the press. The *Toronto Star* sent a reporter, Fred Griffin, to New Jersey, and the stories he filed of the flight north and west were picked up by dozens of newspapers. Before the journey began,

---

[31] *Toronto Star*, April 5, 1926.

a front-page article in the *Toronto Star* on March 18 described it in some detail. "STAR MAN IS PASSENGER ON LONG WINTER FLIGHT, NEW YORK TO RED LAKE," trumpeted the headline. The article reads, "The Curtiss-Lark biplane is to leave Long Island tomorrow to Red Lake via Toronto and Hudson . . . the trip, about 1700 miles long, will be one of the longest winter flights on record, most of it over bush country . . . it is only about 200 miles short of the distance between Newfoundland and Ireland."

Bad weather prevented the plane's on schedule takeoff, but four days later the first stage of the long flight was made. This trip, from Long Island to Buffalo, was described by Griffin as "like a ride on cobblestones." Three days later,as the plane was again stalled by bad weather, the next hop was completed. "RED LAKE AIRPLANE ARRIVES AT LEASIDE; Low Flying Necessary On Journey From Buffalo This Morning" was the headline.

Griffin explained, in his article, that the route which had originally been planned for the long journey was being changed because of the spring break-up beginning in the north. The Lark, at this point, had its wheels on, but it would soon be leaving the developed (and warmer) south, where there were clear paved runways suitable for wheels. A mid-route changeover place had to be located where there was only a little snow on the ice surface: smooth enough for a landing to be effected with wheels, but covered with sufficient snow for the plane to be able to take off again after the wheels had been changed to skis.

It was decided that Sudbury would be the spot where the wheel-to-ski change would be made, and Sudbury was reached on the 29th of March. There was a large sigh of relief, then, as it seemed all would be well and Hudson would be reached in no time. (In fact, it took another 14 days.) "RACE AGAINST DARKNESS BRINGS LARK TO SUDBURY; PLAIN SAILING AHEAD NOW" proclaimed the page one headline of the *Star*. Now that they had finally reached Sudbury, explained Griffin, they were in "the real north, where men are men and ice is ice"; and the journey should go quickly, no longer held up by "the soft south, bad weather, clearance difficulties and the public accounts committee."

Unfortunately, bad weather was not only found in the soft south. The Lark was delayed in Sudbury three days by blizzards, after it had been prepared to go, and then took off in a slight break in the weather. The snow and wind began again shortly after takeoff from Sudbury, and the plane made two forced landings. The second was on Pogomasing Lake, and people who lived in the nearby lumbering settlement of Wye saw the plane land and brought the men home for the night.

On April 7, the plane got almost to Chapleau. "Plane Is Now Near Chapleau, After Battling Very Bad Gale–The Going Is Slow" read the *Star's* headline that day. Griffin, sounding somewhat discouraged, remarked in his article that "bad weather seems to dog our flight into Red Lake. Indeed considering the season . . . it will be considerable of an achievement from a flying point of view to bring the ship up to the gold camp at all."

There were more spring storms during the next stages: to Amyot, Ontario, to Orient Bay, on Lake Nipigon, and finally to Sioux Lookout on April 11. There the pilot and reporter were guests of honour at a dinner with the mayor and other local dignitaries. "LARK IS ON THE THRESHOLD OF RED LAKE'S WEALTH", proclaimed the headlines. And when they finally reached Hudson, April 12, Harry Oaks took over as pilot and flew the Lark to Red Lake. The Lark had carried a number of letters from Toronto and they were all back-stamped, on arrival, "RED LAKE, APRIL 12, 1926."

The first arrival of a commercial flight to Red Lake in early March had coincided almost exactly with the opening of the Mining Recorder's office at Red Lake, an equally momentous event in the primitive camp. Bert Holland opened the Mining Recorder's tent for business, on March 1, and Neh Faulkenham was given the privilege of recording the first three claims in the brand-new mining division. From March 1 to March 4, 371 claims were recorded. From March 21 to April 21, over 3500 claims were recorded. The one-day record was 163 claims. During the most hectic time, the Recording Office was open 24 hours a day, as the eager prospectors, who had rushed in to Red

Lake to stake a piece or two of ground, rushed over to the big tent near the Dome tents to record it.

The recording of claims, however, was not without its problems (or its lighter side, depending upon the point of view). The recording was usually done on pre-printed claim forms but–due either to underestimation of the number of claims that would be recorded, or to the fact that the forms were lost in a load that broke down on the way–the office ran out of forms almost immediately. One result of this was that the very first piece of mail to go out of Red Lake by air was Bert Holland's urgent request for more forms.

Another result was that most of the eager stakers had to write out their own application forms. This process was described in some detail by Gordon Shearn, who mushed in to Red Lake in March to stake claims for Major Cunningham-Dunlop. "Upon asking for application blanks . . . Bert Holland would hand [the person] an application blank and a flock of foolscap with instructions to return the application form for someone else to copy from. If you have never witnessed a grizzled old bushwhacker trying to copy all the printing on an application form with a carpenter's pencil (which was what most of them used for staking), then you have certainly missed something . . . I retired with a bundle of foolscap and, thank heaven, not a marking pencil but a writing pencil. It was at this bright moment that I learned that the boss had a total of 84 claims, more than 60 of which were not recorded, and further that the rest of our party could not do much more with a pencil than sign their own name. The next twenty-four hours were very busy for me."[32]

The lack of printed application forms inspired at least one man to turn a dollar from the situation. One such man was George Punker, who was working on a Great Lakes freighter when he read of the Red Lake gold rush in a Cleveland newspaper. He decided this was for him, and left immediately for Winnipeg. There he found a partner, took the train to Hudson,

----

[32] Gordon Shearn, "Early History of Red Lake," *Precambrian*, April 1956, p. 12.

bought supplies and a toboggan and pulled it to Red Lake himself in 14 days.

Among the supplies was a typewriter, on which the versatile Punker intended to write articles for the *Chicago Tribune.* (It is not known how he intended to get his stories to the newspaper.) The typewriter, however, turned out to be useful in another way. When the Recorder's Office ran out of printed forms George set up his typewriter in an empty tent and typed out forms for recording claims at top speed, making three carbon copies at a time. They sold, for a dollar each, faster than he could type them. As Gordon Shearn remarked (Shearn was established temporarily in the same tent as Punker, writing out 60 application forms in longhand), Punker "made real wages for a while". He did so, in fact, until this small business enterprise was ended by the arrival of a typist and typewriter, loaned by Dome Mines to the Recorder's Office.

In addition to his duties as Mining Recorder, Holland was also appointed Justice of the Peace for the newly inhabited territory. (This was announced in the *Toronto Star* of March 13 as, MINING RECORDER AND J.P. UPHOLDS ORDER IN AREA AS LARGE AS ERIN.) This additional responsibility does not seem to have demanded much of Holland's time, with the exception of attending to the odd case of bootlegging or conspicuous consumption of alcohol. It was fortunate that his second official function did not prove as onerous as his first certainly was.

There was a constant stream of prospectors into Red Lake in March. It was announced on March 1 that, because of the unprecedented volume of traffic to Hudson (and hence to Red Lake) on the CN line, henceforth Hudson would be a regular stop for all CN trains, not just the local trains. One group of prospectors reported that on their trip back out from Red Lake to Hudson they had encountered 53 dog teams, 2 horse-drawn toboggans, and 3 prospectors on foot pulling their own toboggans. In addition to the well-travelled route to Red Lake from Hudson, prospectors were also arriving at the scene of the gold rush via the Bloodvein River from Lake Winnipeg, and from Norway House at the north end of Lake Winnipeg.

The men flooding in to Red Lake were of all types. The majority were the "grizzled old prospectors" mentioned by Gordon Shearn. These were people like F. C. Bigood, whose search for minerals had taken him to Sudbury before the Boer War, to the Yukon, British Columbia, Montana and Idaho. Bigood told a reporter that the present stampede was the only thing he'd seen to compare with the Klondike rush.

There were many men on the trail to Red Lake who had been to the Klondike almost 30 years before, and one report in the press said enthusiastically that "RED LAKE IS MAGNET TO KLONDIKE VETERANS". This report was based on a conversation with one such veteran, Jack Lunney from Kimberly, B.C., who said that he knew of "at least a hundred prospectors from Cranbrook and another 50 from Fernie who are already preparing to leave for Hudson."[33]

Another veteran prospector on the trail, S. C. King, had only one good arm and a hook on his other, mutilated limb, but told a reporter he was managing very well–he could hook the coffee pot right out of the fire when others would burn their hands doing so. A typical old sourdough at Red Lake was Tim Benbow, originally of Timmins. "VETERAN TRAVELS FROM MEXICO TO TEST FORTUNE IN RED LAKE: Prospectors in California and Across Rio Grande More Excited Than Canadians in Ontario About Red Lake", extolled the headline in the *Toronto Star* of March 16, after an interview with him. Benbow had hurried up to Red Lake from a placer gold rush in southern Mexico when he heard the news, and explained to a reporter, "I was in San Dominic . . . and old prospectors . . . were calling me up at all hours of the day and night to ask what kind of an outfit they would need up here . . . "

There were other prospectors who migrated to Red Lake from the placer gold fields of the southwestern U.S.A. And in addition to the old veterans with prospecting experience in Canada and the U.S., there were people from Danzig, from South Africa, from Australia and New Zealand, some of whom stayed in northern Ontario when the rush was over and eventually

[33] *Toronto Star*, March 15, 1926.

made their homes in Red Lake.

One of the Canadians who stayed on at Red Lake was Cliff Harvey. Harvey, one of ten children, was born in the Ottawa Valley, and left home at the age of seventeen. He went west several times on the harvest excursions, which were common in those days, and at the age of 21 decided to try his hand at prospecting. He got his first miner's license then, and spent the rest of his life–up to his death in 1982–searching for the big find. (And living a very independent outdoor life.) Harvey was prospecting in the Porcupine district when he read about the Red Lake gold rush in a newspaper and took the train to Hudson. There he met Gerald Rowan, en route to Red Lake to stake claims for a small syndicate, who hired Harvey to help him. Finding everything staked at Red Lake, the party went south to Gullrock Lake to stake claims.

After breakup Harvey built himself a cabin at Red Lake to work on his claims, and he lived in cabins there for 52 years. Besides working on his claims (none of which ever produced anything), Harvey worked occasionally for other people and made a few dollars trapping during the winters (though he said in an interview in 1980 that the last good year for trapping was 1928). The last cabin Harvey lived in was a rough structure on an island close to the town of Red Lake. With his long, white beard and benign weatherbeaten face, Harvey looked like the archetypal prospector.

Not everyone who went to Red Lake during the gold rush was really after gold. One such was George Johnson, who was offered $100 by a prospector for a ride to Red Lake on Johnson's sled. Johnson refused because his cargo was particularly fragile, a full load of contraband bottles of whiskey. Johnson was one of many like-minded individuals who took bottles of liquor to Red Lake, hidden in four-gallon gasoline cans, by dogsled.

One of the many city-slickers who mushed in to Red Lake was David Wilson, a young Irishman who was working in a hospital in New Jersey when he read of the gold rush in a newspaper. He immediately took the train to Hudson, disembarked in his city outfit of suit and overcoat, oxfords and overshoes, and, in that apparel, walked to Red Lake. Finding everything staked in Red Lake, he walked back to Pine Ridge, then east to Woman Lake where he staked some claims, then back to Pine Ridge just before break-up. The total distance he covered in his overcoat and overshoes was about 285 miles. He lived at Pine Ridge for the next few years, where he was distinguished by his curly black hair that he let grow to his shoulders. (His long hair, apparently, was the result of an altercation with a barber who refused to cut Wilson's hair on credit.)

A pair of young adventurers whose trip to Red Lake was described by the press in some detail, were two young Finns, Toivo Pulkkinnen and Emil Oja. Toivo, a sturdy young man of 21, could not find work at Winnipeg and, reading of the gold rush at Red Lake, decided to head in that direction. Having no money, he walked along the railway tracks from Winnipeg to Hudson, and was joined by a similarly-minded companion en route.

He described his trip to a reporter in Hudson in matter-of-fact terms. "We got a few things to eat at houses and camps along the way, slept where we could and walked when we were too cold to sleep."[34] Toivo was fortunate enough to meet Jack Hammell shortly after arriving at Hudson, and Hammell, who was noted for his kindness as well as his business successes, agreed to put Toivo and his friend on the payroll of the Howey Syndicate and to advance them money. And so Toivo and his friend left for Red Lake, still walking. A front-page selection of photographs of the Red Lake rush in the *Toronto Star* of March 11 featured Toivo carrying a sign: UNEMPLOYED/ WALKING/ WINNIPEG TO RED LAKE 6 March 1926.

Things were going very well for Jack Hammell at this point. Dome Mines made a further announcement on March 18 about their agreement with the Howeys and the Red Lake Prospectors' Syndicate, to the effect that they were undertaking an option to buy 75% of the claims of the Red Lake Prospectors' Syndicate for $500,000. (The *Northern Miner* did note, however, that this deal was simply an exploratory one, and did not commit Dome to much actual capital investment unless the diamond-drilling

[34] *Toronto Star*, March 8, 1926.

results actually warranted it.)

And although Dome was the biggest firm doing exploratory work at that stage, there was a long list that daily grew longer of syndicates and corporations involved at Red Lake. These ranged from simple prospectors' syndicates, like the small group that originally backed Lorne Howey, to large mining firms like McIntyre-Porcupine Mines. There was the Foley Red Lake Mining Company, staking claims in Trout Bay; the Abate Red Lake Development Company, at Golden Arm Bay; the Red Lake Development Company, whose claims were being worked by Pete McDonough and "Klondike Bill" Woodney; there were the Keewatin-Porphyry Development Company of New Liskeard, the British Red Lake Mining Company, the Canadian Red Lake Mining Company, the Hudson Red Lake Development Company, the Ryder and Company Syndicate, the Telluride Syndicate–the list carries on and on.

Money was changing hands in quantity, both before and after shares hit the stock market. Major Cunningham-Dunlop's 30 claims around the Chukuni River and Balmer Lake were incorporated as Dunlop Red Lake Mines, and shares were offered for $30 each. Bobby Cockeram sold 3 claim groups of 9 claims each for $15,000 per group to two Toronto firms who planned to form two new syndicates and to sell shares. The Miller Red Lake Syndicate was selling shares for $25 each; the Evans Red Lake Syndicate offered 50,000 units at 40 cents; the Red Lake Centre Syndicate (whose claims adjoined the Howey's) offered 300,000 units at $4.00; the Red Lake Consolidated shares were offered at $100, the Porcupine Red Lake Syndicate Shares at $50. The business and stock market pages of the Toronto newspapers, the biggest indicators of mining activity in Canada, were dominated, in March, by Red Lake.

There were surprisingly few arguments over stakings that occurred in the hasty and disorganized rush to claim property rights at Red Lake. Most potential conflicts seem to have been settled in the manner of Bill Tyrell. Bill was the son of the engineer who had surveyed the claims of the North Western Ontario Development Company at Slate Bay in 1897. In March of 1926, Bill flew to Red Lake with a companion named Ben Fairwell to restake those old claims. (Fairwell is remembered by pilot Harold Farrington as one of the few passengers who disregarded his advice to keep his baggage under the seat. Fairwell opted instead to hold his bag of mining maps and medicine tightly on his lap. On the flight to Red Lake he watched in horror as the plane hit an air pocket, dropped suddenly, and the precious bag flew out of the open cockpit into oblivion.)

When Tyrrell and Fairwell arrived at Slate Bay, they found the old survey lines and started restaking the old claims. As they were running their claim line southwards, they met another prospector who was running his northwards. Each party stopped where they met, and ran its east-west line from that point. (And that was why, as Tyrrell explained later, the bottom row of claims of Tyrrell Red Lake Mines was only half the normal size.)

The only known case where a serious dispute over claims ever reached the courts was at Pipestone Bay, the most northwesterly extension of Red Lake. In March of 1926, John Y. Cole, Jr. found visible gold in quartz near Pipestone Bay. He staked the maximum number of 9 claims and later these, with another 42 claims, became the property of Cole Gold Mines. (This company, as it turned out, never progressed beyond sinking a 550′ shaft and doing some development work.)

Also in March of 1926, Gus Dupont of Mattawa came to Red Lake to stake claims. Dupont was financed by a large American brokerage firm, and he and his party staked some 50 claims in the Pipestone Bay area. They also did the first year's assessment and development work on them.

The exact sequence of events that followed these stakings is uncertain. It is known, though, that serious disputes developed over the claim lines. (Some local wag, reading in the newspaper that fighter Jack Dempsey had a shortage of sparring partners, is reported to have remarked that Dempsey should come up to Pipestone Bay, as there was no such shortage there.) In any case, litigation ensued and the legal costs eventually ran to well over $100,000 (a substantial sum in those days–and $100,000 more than the Pipestone Bay claims ever produced in gold). John Cole is still spoken of in Red Lake as something of a swindler.

One of the best-documented journeys to Red Lake in March of 1926 was made by Gordon Shearn, who was to spend the rest of his life at Red Lake and to help to develop Balmer Township.[35] Unlike most of the hopefuls who went in over the trail to Red Lake, Shearn's motive was to find, not gold, but work. When he arrived at Hudson in March, the first thing that caught his attention (in addition to "the bustling mass of humanity", as he called it) was the very large number of dogs. Many of the gold-seekers who got off every train were, Shearn remarked, "suffering from a bout with John Barleycorn, for although Ontario was still dry, Winnipeg was very wet. On most of the trains some joker had chalked 'Red Lake Or Bust'. With the men, of course, came more dogs and the resultant din of dog-fights, drunks, howling dogs, whistling trains and roaring planes, I'm sure made Hudson for a few weeks the noisiest town on earth."

In Hudson at that time there were three places to sleep: Kenneally's hotel (which "cost real money"), the Indian House of the Hudson's Bay Company (which cost a dollar a night), and the railway station waiting room (which was free). Shearn opted for a small space on the floor of the station. "What sleep I had", he said, "was definitely terminated by 4 a.m. when various parties desirous of making an early start decided to trek outside to cook breakfast. Although I was near a corner of the waiting room, it seemed no-one was able to get out without walking on me, so I decided to get a breath of fresh air myself, not a bad idea as can easily be imagined although it was plenty cold outside. Besides the hotel dining room there were two buildings which had hastily been converted into restaurants, although many of the prospectors did their own cooking in the great outdoors."

After breakfast, Shearn strolled down to the ice of Lac Seul to look at a plane there–one of Jack Elliott's Jennys–which reminded him of the planes he had seen in France when he

----

[35] I am indebted for the following account of Gordon Shearn's first journey to Red Lake to his article, "Early History of Red Lake," in *Precambrian*, April 1956, pp. 9-14.

served with the British Army there in World War I. When he found out that it would cost around $200 for him to fly to Red Lake in the "crate", Shearn decided that the cheaper methods were for him.

Later in the day Shearn was approached by a stranger who asked him if he had ever driven a dog team. Shearn replied that he had (though he neglected to add that it was only once) and was hired on the spot. Shearn's new employer wanted two men, so Shearn obligingly found a partner in less than an hour–"a big lad straight from a Manitoba farm whose chief attraction for me was a long heavy overcoat. The only bedding I possessed was a pair of lumberjack blankets decidedly inadequate for sleeping in the bush, but with that overcoat between us we could make out quite well."

Agreements were signed (Shearn and his partner were to receive $75 monthly, plus board, and could each stake one claim), the unfortunate-looking team of dogs duly arrived, and after loading all the gear (including a 16′ canoe) on the toboggan, Shearn and his partner retired to the station waiting room for the night.

When they hit the trail early the next morning, they found it easy to follow–in spite of the fact that Shearn and his partner were both "green as green paint, without a compass or a map between us". When they came to a fork, said Shearn, they simply took the more travelled branch, and never had to remain long in doubt as they soon met another party, either going in or coming out. And the trail was also well-marked by what Shearn euphemistically referred to as "painkiller bottles"–as he said, there must have been hundreds of "short snorts" taken on that trail.

According to Shearn, the rush over the trail to Red Lake was, in general, courteous and orderly. "The crowd seemed to regard one another as neighbors and while of necessity there was hustle and enthusiasm it was governed by common-sense, and to my knowledge no-one ever found himself in trouble without also finding many willing hands to help him on his way. . . . [Similarly] on being overtaken by a faster team the man ahead would pull over at the first opportunity and allow the faster team to go ahead." The only fights Shearn saw, in fact, were dog fights, which occurred with some frequency, and "the settling of which required the use of both hands, both feet, and anything one could grab in the way of a lethal weapon."

The first day's travelling brought Shearn and his partner to the Lac Seul Hudson's Bay post, and the second, about 25 miles up Lac Seul, to Sand's Fish Camp. The next day they reached the Hudson's Bay post at Pine Ridge, where "the traffic thinned somewhat as many took the trail north[-east] to the new field of Woman Lake." In fact, at Pine Ridge Shearn met Tommy Powell, whose gold find at Woman Lake sparked most of the interest in that area (an interest which was to eclipse later in the summer the attention paid to Red Lake, both in the press and on the trail).

When Shearn and his partner finally arrived at Red Lake, "having hiked every step of the way as the load was far too heavy to permit of us riding", they found a community of tents, "hundreds of them pitched cheek to jowl . . . all sizes, shapes and colors. At all hours of the day some party would be striking their tent to pull out and the idea was to grab their tent-site with the bedding brush already cut if you were quick enough."

On the last lap of the trail to Red Lake, Shearn had met a number of Indians with dog teams, either going out from Red Lake with empty toboggans or going in, heavily-laden. This was unusual, as the Indians usually acted as guides or mushers for others' dog teams, but rarely drove their own. The reason for this, Shearn discovered, was that the Hudson's Bay post at Red Lake had been rapidly depleted of food supplies, and the Indians had been hastily hired to bring in fresh supplies from the Pine Ridge post, at $7 per hundred pounds.

While the price of food at Red Lake during the gold rush never reached the levels of the Klondike rush thirty years earlier, it did nevertheless follow the laws of supply and demand, to rise to levels anywhere between 4 and 10 times Winnipeg or Toronto prices. The food supply at Red Lake reached its lowest point in April, just before the arrival of the first 18 teams of horses with freight and supplies; at that point there was not much food around except salt pork, bannock, and the odd tin of canned goods. There was a grave shortage of corn meal for dogs from mid-February until after break-up, but there was never any serious shortage of other food. This was principally because the Hudson's Bay post at Red Lake was able to replenish its stocks from the post at Pine Ridge by the use of a bevy of Indian drivers such as Shearn had encountered–and the Pine Ridge post, in turn, was able to get supplies of food sent by horse-drawn sleighs from Hudson to Pine Ridge, loaded to the hilt.

It seems that prices rose quickly as the food travelled on the last lap of its journey, from Pine Ridge to Red Lake. Bacon that cost 60 cents a pound at Pine Ridge, for example, cost $1.25 at Red Lake. While both canned goods and fresh fruit and vegetables were often scarce at Red Lake, there were always flour, salt pork, beans, bacon, tea, powdered potatoes, rice: the basic staples of the prospector's diet. Any variation in food available was rapidly bought up: one 19-year-old, who went in to Red Lake with $2 worth of food for himself, rapidly sold it for $12.00.

It is interesting to compare prices at Red Lake with those in Toronto and Winnipeg at the time. A hundred-pound bag of sugar that cost $6.00 in Toronto and $6.95 at Winnipeg cost between $25 and $35 at Red Lake. A hundred-pound bag of flour cost about $6.00 in Toronto and $4.50 in Winnipeg, but jumped to $14.00 at Red Lake. Other prices at Red Lake were $1.50 a pound for butter (47 cents a pound in Winnipeg), $1.35 a pound for tea (66 cents in Winnipeg) and $1.00 a pound for lard.

These prices should be seen in the light of the standard wages in the area for labour at that time: around $3.00 per day, plus board. Another indicator of how much it cost to eat at Red Lake can be gained from comparing these prices to the costs of food today, and then to the costs of other basic commodities at that time. In the spring of 1926, in the cities of Canada, you could buy a pair of men's tweed trousers for $2.29; 3 pairs of good-quality towels for $1.29; a new 7-room brick house in North Toronto for $6,000, or a new 5-room bungalow in the River Heights section of Winnipeg for $5,000.

The prices of food were higher at the Klondike, 30 years earlier, than they were at Red Lake, but there was a major difference. In the Klondike rush, many men were lucky enough to find nuggets of gold that made all their expenses insignificant. But that didn't happen to anyone during the Red Lake rush.

The Klondike gold rush was certainly the one to compare to Red Lake in the eyes of the American press, at any rate. The comic strip "Mutt and Jeff" on March 25, 1926, had its heroes set at Red Lake. Mutt and Jeff were busy panning free gold (which, of course, never happened at Red Lake because the gold was not of this kind). To their delight, they found $20 worth of gold–but to their chagrin, they discovered when they hit the local restaurant that a square meal cost $30! (This scene embodies more than just the misconception about panning gold. At that point there were no restaurants at Red Lake, although the Dome Mines cook tent sometimes served meals to non-company personnel.)

The new territory at Red Lake continued to receive additional flickers of recognition from the government and institutions of Canada. Bill Brown's modest fact-finding mission for the postal service was followed by a trip to Red Lake in March by R.H. McNab, General Manager of Railway Mail Services. McNab announced after his trip to Red Lake that a "temporary mail service" to Red Lake, once every two weeks by dog team, had been instigated. Bill Brown was appointed Mail Carrier to Red Lake on this bi-weekly trip, and the toboggan on which his dog-team hauled the mail was emblazoned with the Royal Cipher and the words, "Royal Mail."

The almost complete lack of any medical facilities for the men streaming in to Hudson and taking off from there for Red Lake, had also received official notice. The Ontario Division of the Red Cross had been formed right after the end of World War I, a war in which the Red Cross had played a major role in caring for the wounded and the dispossessed. The newly-created Ontario Division found that health services, in the large part of the province lying beyond the settled south, were virtually non-existent. A report stated that large areas of the thinly populated hinterland had been almost entirely overlooked in terms of health care; there were few, if any, nurses and not enough doctors; many areas had no hospital within 100 miles.

In 1922, then, the Ontario Division had decided to take on the formidable task of supplying "the vast outlying Districts of the Province with Outpost Hospitals and Health Centres."[36] The first outpost hospital was opened at Wilberforce, in Haliburton. And, after the terrible forest fires that devastated the Haileybury area in 1922, when the Red Cross dealt with the emergency, outpost hospitals were established at both Haileybury and Englehart. The first Red Cross outpost in northwestern Ontario was opened at Dryden in 1923, and it was rapidly followed by several others: at Nakina, Rainy River, Hornepayne, Quibell.

As the prospectors began to pour off the trains at Hudson in early 1926, a committee of the Red Cross dealt speedily with the suggestion that medical services were needed there. A railway car was borrowed from the CNR, parked on a siding at Hudson, and outfitted as an emergency nursing station. The outpost station opened for business on March 1st, with two nurses in charge. The nursing station treated some 500 patients in its seven months of operation. Among the first was an unfortunate chap who mistook a loose sleigh dog for one of his own, and received a badly bitten hand. Among the most dramatically-injured patients were pilot Howard Watt and his passenger after their plane crashed on the ice of Lac Seul on March 27. The clientele at the nursing station also included a number of gold-seekers with pneumonia, induced by sleeping on the snow without enough blankets.

The establishment of the Red Cross nursing station at Hudson was covered by the press in minute detail. March was the month of most intense press coverage of the gold rush to Red Lake–this was both a cause and an effect of the parallel fact that March was the month of most intense activity on the trail to Red Lake.

----

[36] E.H.A. Watson, *History, Ontario Red Cross, 1914-1946*, Toronto, 1947, p. 14.

Reports of activity on the gold rush scene were featured almost daily in the Winnipeg and Toronto papers (in Toronto, they were usually front-page news), and very frequently in all the other Canadian city daily newspapers as well. Old stories were hashed and rehashed to provide new copy for the eager reading public. A prime example of the search for news about this media event (and the resulting printed proliferation of non-news) was an article in the *Toronto Star* on March 25. The headline of this piece read "HUDSON ARCHER HASN'T GOT FEVER OF THE RUSH FOR GOLD", and the subject of the article was one N.F. Gale, an employee of the CNR who lived at Hudson, whose hobby was archery, but who wasn't interested in going to Red Lake.

The international press coverage was extensive. Major Cunningham-Dunlop's frequent conversations with the *Toronto Star* reporters were picked up by many American newspapers, and as a result he received a large number of letters from every corner of the U.S. The obliging Major forwarded a number of these to the *Star*. The letters he forwarded were from Spokane, Washington; from Montana; from Michigan; from Duluth; from Indianapolis. One that was quoted by the *Star* was from a school-teacher who signed herself as head of a committee of like-minded teachers in Grand Marais, Minnesota. She asked the Major if the claims at Red Lake were all for sale, if they varied in size and price, and to whom she should write to buy one. Her letter ends on a note of some pathos, as she said, "We have caught the [gold] fever. . . . None of us is wealthy. So far we have just done a little better than to break even when the school year [ends] . Therefore we [are] interested in any wild prospect which might leave us more independent in our advancing years."

The London [England] *Times* of March 16 carried an article entitled "AN ONTARIO GOLD 'RUSH': THE RED LAKE CAMP", which ended with the optimistic words, "The prediction is made that the 'rush' into Red Lake will rival in importance that which followed the news of the rich finds in the Cobalt and Porcupine districts of Ontario." The article is accompanied by a map, and elsewhere in the edition are two photos taken at Hudson. The *New York Times* carried several articles on the gold rush during the month of March; the first, on March 6, covered the establishment of Jack Elliott's air service from Hudson to Red Lake.

The second *New York Times* article in March, which appeared on the 12th, was based on a rather extraordinary interview with Charles Henrick. Henrick was the prospector from Englehart who had gone in to Red Lake with the Dome party at the beginning of January. Henrick, described by the *Times* as an "ex-soldier prospector", eventually settled in Red Lake.

His descriptions of the "picturesque features of the trip" are bizarre in the extreme. The article says that the "Red Lake country has a population of tribes of Indians unknown to Henrick, and . . . the men are discovered to be wearing serge suits and fancy colored or pleated shirts, without overcoats of any kind, even though the temperature was 55 degrees below zero when Henrick was there." (The temperature at Red Lake has never been recorded as low as this figure, and all extant photographs of Indians who participated in the gold rush as drivers or mushers show them to be wearing relatively normal clothing for the climate.) The article goes on to say that "the Indian will not take serviceable clothes from the Hudson's Bay posts. He is a simple child of the forest and is laughing almost constantly. . . . His coat is covered with colored buttons and pins."

Less fanciful, and much more wide-ranging, was the feature article in the Sunday *New York Times* of March 28. This extensive and well-illustrated article, in the prestigious *Times*, represents the peak of international press coverage of the gold rush. The author caught very well the old/new contradictions of the rush. The piece was headlined, "AGAIN GOLD LURES MAN NORTH: But the Newest Treasure Trove in Canada Is Different From Others Because the Airplane, Snow Tractors and the Radio Play a Part in the Dash Into the Remote Ontario Wilds."

The side-by-side existence of dog teams with snow tractors and airplanes, seized upon by the journalist as the dominant feature of the gold rush, was in fact unique in the history of those strange mass movements to find gold. This kind of contradiction extended to the settlement at Red Lake, too, in the journalist's description of the tent city there; "The Red Lake Goldfields presents some of the most striking contrasts imaginable, for here is a mining camp so primitive that the miners have only their tents and their fur-lined parkas to sleep in. . . . Yet during the daylight the air is whipped by the propellors of airplanes carrying in supplies from the railway and the roar of motor tractors."

The correspondent's view of the country was not a flattering one. He refers to the "sub-zero Arctic gales that sweep through this Canadian no-man's land", and later expands on this theme by reflecting, "It's funny what gold will do for a place. This stretch of desolate lake country, mosquito-ridden in Summer, rain-drenched in Spring and Fall, almost snowbound in Winter, would seem . . . just about the least attractive place on the continent . . . "

But the rest of the article rings with the notes of fulsome optimism that are typical of the press coverage of the gold rush. "Through [the rocks at Red Lake] run veins of gold, so rich that the mining world . . . was . . . pinching itself for conviction that it was not dreaming." Referring to Red Lake as "the newest Eldorado", the author says that "To two brothers, Lorne and Ray Howey, is conceded the honor of first striking gold at Red Lake. They are now . . . satisfied in knowing that the claims they have staked will yield them far more dollars than they will ever need . . . "

To this correspondent, the mushrooming village of Hudson typified the boom spirit. "Hudson consisted of a station house, a general store and a cabin a few weeks ago. Today it resembles the typical frontier town suddenly overtaken by a sensational mining boom. . . . New buildings are rising, and the streets are lively with the throngs of prospectors, dog drivers, packers, Ojibwa Indians, camp followers. Every incoming train adds to their numbers."

The feature article ends on a high note. "The pessimist about Red Lake has yet to be heard from. . . . C.E. Holland, the Government agent, has already recorded 4,000 claims . . . [Holland told the reporter that] 'a lot of us are figuring on a railway next year. It will be worth it if the camp pans out like everyone thinks it will.' . . . In short, as a young civil engineer from New York remarked, 'This Red Lake show looks like another Klondike . . . only with up-to-date trimmings.'"

Transportation, whether by dogsled or airplane or by any of the methods between these two extremes on the technological scale, remained one of the central concerns of the gold rush. Dogs–hungry, hard-working, snarling, fighting, and generally well-cared-for–remained the most common source of motive power. There were many reported cases of dogs giving birth to litters en route, with the newborns either being bundled in a blanket on the toboggan, or left in one of the commonly-used shacks along the trail for the next prospector to tend.

At the Dome camp in Red Lake, John Hammell was nursing half a dozen unhappy huskies whose shoulders were rubbed raw from their collars. He jokingly referred to his row of kennels as "The Red Lake Humane Society", and told a reporter that most of the dogs wouldn't be fit for work again that winter and that the camp was short of dog food, but he didn't have the heart to shoot them. (The dogs, while extremely useful, were not irreplaceable: there was more than one bizarre party noted on the trail, in which an unfortunate fellow had lost his lead dog, and harnessed himself up at the head of the pack to take the missing animal's place.)

Other methods of transportation were still being attempted. William Loranger, from Quebec, had built a primitive snowmobile in 1923 and, in early 1926, he shipped it to Hudson to transport him up to Red Lake. The machine, unfortunately, went to Hudson, Quebec instead and Loranger went by dog team to Red Lake, where he staked some claims. When the snowmobile arrived at its proper destination, break-up had begun and it was too late to use it; however, Loranger did use the machine around Hudson in the following winter.

There was still hopeful speculation about a road to Red Lake, particularly in the Kenora area. The Board of Trade, there, had hopes of persuading the Ontario government to extend the 20-mile road to Redditt (north of Kenora, on the CN line), north to Red Lake–hopes that proved to be in vain.

The main focus of anticipation, though, was a railway to Red Lake. This also was a source of some controversy, all quite futile, as it turned out, as no railway was built. But in March of 1926 the word railway was on everyone's lips. Who was to build the railway to Red Lake, and from where? The *Manitoba Free Press* carried a headline on March 1st that read "WINNIPEG WILL BE HEADQUARTERS OF NEW GOLD FIND". This would happen, the journalist explained, if power and rail connections were made to Red Lake through Manitoba. Neither of these types of connection were, it was admitted, presently in the planning stage, but if they were built, they should extend from as yet unbuilt services to the mining areas in central Manitoba.

An editorial a few days later in the *Free Press* reiterated the theme that "These areas, whether in Manitoba or Ontario, are a natural treasure territory for Winnipeg and will mean much to Winnipeg and Manitoba." (This Manitoba-centred point of view, naturally enough, was pervasive in the *Free Press's* coverage of the gold rush. One article traced the rush itself back, not to the Howey brothers' discovery, but to a gold find by three Manitoba prospectors in the East Bay area of Red Lake in 1922. In fact, a meeting was organized in Winnipeg in early March to bring together major Winnipeg business interests in a common cause: if they weren't careful, Toronto rather than Winnipeg would become the supply centre for, not only Red Lake, but the mining areas in central and northern Manitoba as well.)

These were fighting words, for Red Lake was seen as a very large and very rich mining area whose development would send huge waves of capital in many directions. The *Toronto Star* seized on these speculations from Manitoba and trumpeted in its own front-page headlines on March 10, "MANITOBA WOULD BUILD RAILWAY FROM WINNIPEG INTO NEWEST GOLD CAMP: Planning To Divert Ontario's Mineral Wealth Westward, Is Belief." The article went on to say that, "The proximity of Red Lake to the boundary of Manitoba now threatens to . . . raise once more the issue of the rights of one province to build a railroad through the territory of another. . . . The city of Winnipeg is anxious . . . for . . . the benefits of a railroad running into Red Lake, and the rumour is going around Toronto that a

Manitoba capitalist discussed this with officials of the Ontario government . . . and was given the cold shoulder. Premier Ferguson and Hon. Chas McRea say they know nothing about this."

A week later two *Star* reporters called on the Manitoba Minister of Public Works, W. R. Chubb, and he confirmed that a railway extension in the direction of Red Lake was presently being considered, and told the reporters there were presently two bills before the Manitoba legislature to grant charters to railways running up to the Ontario-Manitoba border.

The Manitoba interests were not the only ones reported to be considering building a railway to Red Lake. Government officials were receiving frequent submissions both from mining interests and from municipal groups about a railway line to be built north from the northern CN route into Red Lake. The Kenora Board of Trade sent a delegation for this purpose to see Charles McRea in Toronto. In a separate manoeuvre, D'Arcy Scott, an Ottawa lawyer, announced that he had applied to the federal government for charter rights to build what he called the Red Lake and Northwestern Railway north from the CN line to Red Lake, and then west to Le Pas in central Manitoba. (His financial backers for this enterprise were unnamed Montreal businessmen.)

The *Toronto Star,* on March 11, ran a front-page headline, "ASK THORNTON FOR RAILWAY TO RED LAKE." The event concerned was an interview with Sir Henry Thornton, President of the CNR, in which a number of questions, on many subjects, were put to Sir Henry. One of these concerned "the possible extension of the CNR into the Red Lake gold field", and on this subject "Sir Henry was careful to make no promise." A few weeks later, though, an unnamed CN official was quoted as saying, "Just as soon as the drills confirm surface showings, the CNR will start in from Sioux Lookout or Hudson. They'll be first in. It's their country, and again they'll demonstrate that they are the swiftest railway builders in Canada."[37]

----

[37] *Toronto Star,* March 25, 1926.

While all this heated and hypothetical discussion of roads and railroads was occurring, there was a real and equally dramatic episode in transportation history actually happening on the long winter trail to Red Lake.[38] The thorough exploration of a gold-mining property (to say nothing of the development of it) requires heavy machinery, and neither dog-sleds nor the light aircraft operating at that time were at all adequate for the purpose. Dome Mines had a great deal of machinery to bring in to Red Lake, including three big diamond drills. It was decided that this machinery would have to be transported either by snow tractors or by horse teams, and early in February a start was made at constructing a navigable winter road for either tractors or teams.

To make the tractor road on the uncertain and dangerous ice of Lac Seul, attempts were made at rolling the surface in order to mix the surface snow with the underlying slush, thereby forming a solid, hard surface. And to connect the north-west end of Lac Seul with Red Lake, the dog-team portages, which had been used by Indians and fur-traders for 150 years, were to be widened and somewhat evened out. A gang of Indians was cutting this road from Pine Ridge to Pakwash, while Alex Gillies was directing a group working from the other end, from Red Lake to Pakwash.

On the first of March, some four carloads (around 150 tons) of heavy machinery arrived at Hudson and the job began. Three small crawler-equipped tractors had also been shipped to Hudson, and, in the first week of March, they started out, pulling heavy loads of freight on scows behind them. Unfortunately, spring weather started at the same time as the freight did. There had been particularly deep snow that winter which turned to slush in the unusually early thaw, and it was impossible to construct a hard surface on the ice by rolling it. The lake became

----

[38] The details for the following account of Ken McDougall's journey to Red Lake are drawn from a document in the Archives of Ontario, which is the text of an address delivered by K.H. McDougall to a Gyro International Convention in 1952, entitled "Early Red Lake Days."

impassable for the tractors. (These snow tractors worked well enough on hard snow or ice but wallowed in slush, unable to move.) By the 15th of March there were boilers, diamond drills, pipes, rods and other pieces of equipment strung out in a helpless line almost 30 miles long, along the surface of Lac Seul.

At this point, Dome and McIntyre sent representatives to see Ken and Archie McDougall at McDougall's Mills, a hamlet on the CN line, 14 miles east of Sioux Lookout. The McDougall brothers had been involved in the timber/sawmill business there for several years, and when the gold rush to Red Lake began, they decided that the transportation business offered good potential. They took George Wardrope as their partner in the new deal with Dome and McIntyre.

George Wardrope went on to become an alderman in Port Arthur, a member of the Ontario Legislature, and finally Minister of Mines for Ontario. Ken McDougall, a member of a distinguished Ottawa family (his father was Canada's first Auditor-General) had served as a Colonel in World War I, and spent the rest of his life in northwestern Ontario. After his transportation business folded, Ken set up a general store in Red Lake which he operated, at first, from a scow, since the Howey Mine permitted only one officially-sanctioned store on the townsite.

It was in 1932 that Ken, his wife Margaret (an Ottawa girl who had been in the 1919 graduating class of Trinity College, U of T) and their two young daughters moved to Red Lake, and it was their home from then on. Ken operated a general store in Red Lake, was involved in the sawmill business, and was one of Red Lake's most active citizens until his death in 1972. His widow, Margaret, one of the real pioneer women of Red Lake, still lives in their apartment above the now-empty store, overlooking Howey Bay, in the section of Red Lake called McDougallville.

The initial freighting contract that made transportation history during the gold rush was signed on the 21st of March, 1926, with the newly-formed Red Lake Transportation Company (the McDougall Brothers and George Wardrope) agreeing to transport the freight for Dome and McIntyre. Their fee was to be $200 a ton, with the proviso that they were to be paid their costs plus 10% if the actual transport of the goods cost more than $200 per ton, or if (as seemed likely) they were unable to get the freight through.

Motorized vehicles had proven unequal to the task, and it was decided that horse-teams were the answer. Although the contract wasn't signed until March 21, McDougall and company–realizing there was no time at all to lose if that enormous quantity of freight was to reach Red Lake before the lake ice became too dangerous to travel on–had despatched an advance party on the 19th. This consisted of a driver with two teams of horses, a cook with full equipment and supplies, and a couple of tents. Then arrangements were made to ship the necessary horses and equipment to Hudson. A total of 96 horses (48 teams) were shipped to Hudson, 16 to a carload.

When they arrived they were shod for travelling on ice, the sleighs were assembled, racks built, and freight loaded on. The machinery strung out on Lac Seul was to be loaded en route. A crew of about 65 men–teamsters, helpers and cooks–was assembled. Each teamster, in addition to blankets, lantern, axe, chain and rope, was given an emergency ration (usually a case of eggs, slab of bacon, and bag of bread) in case he had bad luck and became separated from the main party and the cook team/tent. And the party had to carry food for not just men but horses. A thirty-ton carload of oats and three carloads of hay were sent to Hudson for this purpose, and for every two loads of Dome and McIntyre freight carried, there had to be one and a half loads of feed for the horses.

All the teams had left Hudson when Neh Faulkenham approached the new freighting company with an additional assignment. Faulkenham had decided to open a general store at Red Lake, and had assembled some 40 tons of goods for the purpose. He offered McDougall and Wardrope $250 per ton, without any proviso, to bring his freight in. The terms were accepted, and another 16 teams of horses were shipped to Hudson, the goods were rapidly loaded and the teams assembled, and they were despatched to Red Lake.

The going was very rough for the first week, as the horses were not much faster than the ill-fated snow tractors at pulling heavy loads through deep slush. But then a happy turn in the weather occurred. After the mild weather had caused water to accumulate on the ice surface, there was a sudden cold snap which froze the whole surface to solid ice. The whole of Lac Seul became an ice road for a short time, and the teamsters made excellent time. Unfortunately, the original marked ice-road was lost in the thaw-freeze sequence, and two of the teamsters lost their way. They reappeared after a week, with their food supplies for both horses and men virtually depleted.

When they reached Pine Ridge and the ice-road was left behind, the route became much rougher. Eighty miles had been covered, with fifty more to go. The mines had not finished their project, begun in February, of widening the old trail. And because of the unusually mild weather, many of the hills on the portages were almost bare of snow. On one of the worst hills, near Pine Ridge, a teamster could not get his team to pull the sleigh up the slope so he had to take the sleigh apart and skid it up the hill, one piece at a time, then reassemble it at the top.

When the first teamsters reached Pakwash Lake, they discovered no passable road at all, so the men had to engage in road-cutting, with the help of work gangs from the Dome and McIntyre operations. Freight goods were cached temporarily at Pakwash Lake and the last lap of the road was cut through by 2 p.m. on the afternoon of April 8th. At this stage the lake ice had begun to disintegrate near the shores, and temporary log bridges and ramps had to be built to get the teams safely to shore. The first teamsters arrived at Red Lake around noon on April 9th. They returned to the cache of goods at Pakwash Lake (which had been pilfered in the interim), and the balance of the freight arrived at Red Lake on April 10th.

The teams and teamsters turned around immediately for Hudson. This journey was much faster, as the sleighs were empty. It was much more dangerous as well, though, and a total of 5 horses died, from drowning or exposure.

The epic journey of that 190 tons of freight to Red Lake was a real achievement. It was just as historically significant as Jack Hammell's decision, in the previous fall, to use airplanes to get preliminary crews and equipment to Red Lake before break-up. Hammell's decision had inaugurated a new use for bush planes and opened a whole new chapter in Canadian aviation. And the McDougall-Wardrope company's hauling of almost 200 tons of freight over virgin bush, for around $200 per ton, also had far-reaching results. It showed that a mine could be developed at that unprecedented distance from the railway, and that the necessary equipment could be brought in at a price low enough to make it feasible.

Harold Farrington preparing to fly to Red Lake, March 3, 1926.

Horses of the Red Lake Transport Company with freight, April, 1926.

Clerk and Mining Recorder at Red Lake, April, 1926.

Dr. Dingwall and nurses en route to Red Lake, June, 1926.

First steam diamond drill at Red Lake, summer, 1926.

First log house at Red Lake, summer, 1926.

Bill Brown and the Red Lake Post Office, summer, 1926.

Red Lake business section, summer, 1927.

Red Lake Post Office with living quarters addition, summer, 1927.

Curtiss Flying Boat, 1928.

Red Lake Laundry and café-hotel, 1928.

Howey Gold Mine, 1930.

Tractor train in slush, Ear Falls, 1926.

"A Three-Horse Town": Red Lake, late 1930s.

The New York Times
Magazine Section
(Copyright, 1926, by The New York Times Company.)
Section 4
NEW YORK, SUNDAY, MARCH 28, 1926 TWENTY-FOUR PAGES

AGAIN GOLD LURES MAN NORTH

But the Newest Treasure Quest in Canada Is Different From Others Because the Airplane, Snow Tractors and the Radio Play a Part in the Dash Into the Remote Ontario Wilds

Photographs are reproduced by permission of the Red Lake Public Library.

The Gold Rush, Part IV: April-May, 1926

April 1926

The climate and the terrain dictated the means of transportation used at every stage of the gold rush to Red Lake. The month of April is the month of spring break-up in that part of Canada–the month when the ice that has provided a reliable surface all winter disintegrates and is eventually swept out and under by the winds of spring. While this process is occurring the water is not yet navigable by boats. In April of 1926, airplanes could not be used during this interim period, either. (The Jennys were equipped only with skis; the Lark could be equipped to land on ice in winter or on water in summer, but there was no year-round landing strip at Red Lake to use when neither ice nor water was consistently available.) As a result of these conditions there was a virtual news black-out during the time of break-up, which lasted, in 1926, from about April 18 (the date of the last flight of the Jenny) to May 12, when Lac Seul was reported free of ice.

The intense level of press coverage of the gold rush had begun to drop off around the beginning of April, in any case. There were several reasons for this. A minor cause was the discovery of gold in the Woman/Birch/Narrow Lakes area (Woman Lake is about 50 air miles east of Red Lake, Birch Lake another 25 miles north-east from there). Another reason was that nothing new was really happening at Red Lake at this time, and there was a limit to the number of stories that could be written (or read) about prospectors, dogs, and Dome's decision on whether the option on the Howey property would be taken up or dropped. And because there was no more real news, no more novelty, it was time for the press to move on to something else.

The open-cockpit Jenny, run by Jack Elliott, continued to make daily flights (weather permitting) from Hudson to Red Lake during the first weeks of April. Howard Watt's crash landing at the end of March had put one of the service's two planes out of commission, and there were long line-ups daily for transportation in the remaining plane. The Jenny made its last flight to Red Lake on April 18, as it could not be adapted to land on water. And, therefore, for almost a week the Lark and the Jenny were both flying to Red Lake, as it was April 12 when the Lark finally arrived from New Jersey after its lengthy and weather-beaten flight.

Both the Lark and the Jenny made a number of mercy flights out from Red Lake. The *Star* reported on April 8 that Harold Farrington flew a prospector named Stemp, who had "taken seriously ill" (the nature of the illness was undisclosed) to Hudson, where he was treated in the Red Cross railway car, and from there taken to the hospital at Sioux Lookout.

Similarly, one day in late April two men found a prospector on the trail near Red Lake who was bleeding from the mouth. They took him to Constable Carr who seems to have possessed considerable diagnostic skills, as he made a correct diagnosis of bleeding stomach ulcers. (Apparently, a steady diet of dried potatoes, common among prospectors, made this condition a frequent one.) In any case, Constable Carr apparently fed the patient some ice he had crushed from the water pail (a primitive version of a more modern treatment of the malady) and the penniless prospector was flown, free of charge, by Captain Oaks, in the Lark, out to Hudson for medical treatment.

The time of open water was fast approaching. Plans were being made in many places for various types of commercial boat service to take the flow of would-be prospectors to Red Lake. More proposals for a railway were being heard as well. Peter Heenan, the M.P. for Kenora-Rainy River, filed a petition to extend the time allowed for construction of a railway north from the CP line at a point near Dryden to Red Lake, and then west and south to Winnipeg. (This charter, to the Kenora and English River railway company, had originally been granted in 1917.) A plan for boat service from Manitoba was reported in the *Toronto Star* of March 31. A report that day said that according to the terms of an agreement signed between the Manitoba Minister of Public Works and a representative "of a large mining syndicate", the syndicate would put into operation a boat service from Lake Winnipeg to Red Lake, providing transportation over the necessary portages.

Large amounts of both time and money were being expended to provide boat service up Lac Seul to Red Lake as soon as the waters were navigable. The Hudson's Bay Company had ordered two new pointer boats (river boats that are pointed at both ends and have a shallow draft) to operate between Hudson and Pine Ridge. The Triangle Fish Company, which had been operating for some years on Lac Seul, had a new passenger-carrying launch being built in the shipyards at Collingwood. And the Lac Seul Transport Company had a new 40-foot launch being constructed at Kenora.

Meanwhile, the surface trail was in better shape than it had been since the gold rush began at Christmas, owing to that unusual thaw-freeze sequence which had been of such help to Ken McDougall's teamsters. The good surface trail came too late, however, to benefit the unfortunate snow tractors: two of them were towed back to Hudson and the third rested until summer at the bottom of 26 feet of icy water in a bay of Lac Seul, where it had broken through the ice. Everything else, though, moved speedily along. Several Model T trucks traversed the new ice road between Pine Ridge and Hudson. (And they sometimes carried unusual loads. When Bill Tyrell was returning from re-staking his father's old claims at Slate Bay, he drove a dog team as far as Pine Ridge, and then rode in one of the trucks to Hudson with his dogs tied to separate stakes in the back of the truck and he, lucky fellow, in the back with them.)

The dog-teams that were covering the trail under their own steam, at this point, made excellent time. And so did the horses. As soon as the McDougall brothers and George Wardrope had completed their freight contract to Red Lake, they began to carry passengers. The "Haywire Express", as the owners called it, operated for about two weeks. It was in effect a stage-wagon operation, with horses, drivers, sleighs, feed for horses and cook-tents for humans stationed at temporary camps every 15 or 20 miles along the route. The Haywire Express charged twenty cents a pound–a rate that applied, in Ken McDougall's words, "without discrimination to baggage and men."

For several weeks in April, then, the trail to Red Lake was alive with a colourful mixture of dog teams, men on foot and on snowshoes, horse teams, abandoned freight, and motor trucks;

while above, the Jenny and then the Lark flew back and forth.

As the end of April approached, the ice became increasingly treacherous and a number of parties were forced to make last-minute changes to their transportation plans. Five McDonough brothers and Marius Madsen had been staking claims in the Birch Lake area in early April, and returned to Pine Ridge on their way back to the railway. (Pine Ridge was the junction point of two routes: one going west to Red Lake, the other northeast to Woman Lake and Birch Lake.) The McDonough party left Pine Ridge, going south, on April 19. Three days later the temperature rose to over 70 degrees fahrenheit and it rained steadily. After breaking through the ice of Lac Seul twice, the group decided it was too dangerous to continue, and took to the shoreline instead. There were no trails, and it took them eleven more days of very hard going before they reached the railway line.

There were others seen on the trail in late April who were obviously prepared to be versatile, as they had canoes tied to their sleds. And there were still others who, faced by the rapid transition from ice to water while en route, opted for the water rather than the tangled bush on shore, even though they were woefully unprepared for such a choice. Two brothers from St. Catharines had walked to Red Lake pulling their toboggan, staked claims in the Pipestone Bay area, and proceeded back out. They were hauling their gear across a short portage between Keg Lake and Gullrock Lake, just south of Red Lake, when they became stranded. The day was mild, a strong wind arose, and the ice went out. With considerable ingenuity, the brothers felled a large birch tree and built themselves a birch-bark canoe, complete with paddles, and made good speed southwards in it.

The southerly end of the Red Lake trail, at Hudson, was booming now. One of the many anomalies of the gold rush to Red Lake was that very little of the "boom" that characterizes gold rushes everywhere–the instant buildings, the commercial centre that develops overnight, the ladies of the evening who quickly move in to ply their trade–happened at Red Lake. Instead, the booming happened at the two junction points of the transportation system to Red Lake: at Hudson and at Pine Ridge. The rush to Red Lake was the only reason for the instant developments at these junctures. But because they were fairly readily accessible to men, machinery and building materials, and Red Lake was not, they boomed while Red Lake remained (with a few log-cabin exceptions) a tent city.

The scene at Hudson was colourfully described in an article in the *Canadian Mining Journal* datelined Hudson, April 10. "Canoes are piled up in front of the Hudson's Bay Company post like cord wood. Power boats and scows are being overhauled on the shore, not 100′ from the main line of the CNR. Caterpillar tractors are hauling timber and stone for the docks. An incessant hammering loudly proclaims the erection of hotels, restaurants, general stores, etc. Around the buildings of the Imperial and Toronto banks an air of calm assurance already prevails–they have been built more than a week. Out on the ice, like a huge grasshopper, is the plane being petted by its pilot into performing a daily flight to and from Red Lake. Watching all these activities with the keenest personal interest is the best conglomeration of sleigh dogs the writer has seen."

Seventy lots at Hudson had been surveyed and put on the market for sale or lease in March, and development had followed rapidly. The Imperial Bank of Canada opened its new branch in Hudson on April 3, and the Bank of Toronto opened its branch the following day. (The latter event, in fact, was announced in the business pages of the *Toronto Star* in a dignified advertisement: "RED LAKE DISTRICT: the Bank of Toronto announces the opening of a branch in Hudson, Ont. for the purpose of providing banking facilities for the Red Lake District.")

Hudson also featured a new cafe, a new restaurant with sleeping accommodations, a new general store, new docks, a new office building housing the OPP, a doctor, a lawyer, and the Elliott-Fairchild air service office. It seems that the Hudson's Bay post was moved to respond to this unaccustomed competition, for an unusual ad appeared in both the Winnipeg and Toronto newspapers. "Red Lake Mineral District: Prospectors and mining men can obtain canoes and supplies from our stores at Hudson and Winnipeg. . . . We operate a transportation system between Hudson and Red Lake and are prepared to accept consignments, large and small."[39]

The Hudson's Bay post at Red Lake was also encountering its first competition at this point. The 40 tons of food and general merchandise that the horses and teamsters of the Red Lake Transport Company had brought to Red Lake for Neh Faulkenham, had arrived. Faulkenham set up a large tent and a smaller adjoining one near the shore of Howey Bay and opened a dual-purpose establishment, Faulkenham's Store and Restaurant. Faulkenham, the sinewy little man with the big moustache, had faith that Red Lake would not forever remain a community of canvas tents. After the waters became navigable in May, he went back to Hudson and bought a large quantity of lumber to sell at Red Lake. He transported all 2000 feet of it himself all the way over the 180-mile summer route to Red Lake, carrying it on his back over the 5 portages.

On the mining front at Red Lake, the big news–non-news as yet–was the diamond-drilling program being carried out by Dome on the Howey property. Diamond drilling had started about a week after the first teamsters galloped into Howey Bay with the first loads of Dome freight. The company's contract called for 5000′ of test holes to be drilled on the Dome-optioned property, with another 4000′ contracted by McIntyre-Porcupine to be drilled on their adjoining claims.

It was appreciated by everyone at Red Lake, from the Dome officials to the prospectors who had staked bush claims in the most unlikely areas, that the success of all their enterprises hinged on the diamond-drilling program. If it showed good gold values, Dome would go on to build a producing mine on the Howey property and Red Lake would become a real gold camp; if not, the whole temporary settlement would probably disappear.

---

[39] *Manitoba Free Press,* April 9, 1926 and *Toronto Star,* April 12, 1926.

Dome's $50,000 payment to the Howey Syndicate, part of the original agreement signed in January, was made in April. It was reported in the *Toronto Star* on April 23, that at a meeting the previous day of the Howey Syndicate, a $7.00 per unit payment was authorized, to disperse the cash payment among stockholders in the syndicate. This, of course, seemed a good omen to market-watchers and to the hopeful, shifting population camped around Red Lake.

Staking was still going on at an intense pace around Red Lake, even though the remaining unstaked land was in more and more dubious area. New syndicates continued to arrive: one of these was a firm called the Anglo-French Company, of London, England, which announced to the press that it had a crew staking claims at Red Lake. The *Toronto Star*, on April 15, noted that the Mining Recorder's office at Red Lake was still working late every night, but added that, although men were still arriving daily along the trail, their numbers were "very thin" compared to a month ago.

This decrease in numbers was partially due to the fast approaching spring break-up of the lake ice. Prospectors were being warned in the press, by both government and CNR officials, that this was a dangerous time to set out from Hudson. But it was also a precursor of things to come.

Another ill omen for Red Lake was the discovery by Tommy Powell, a prospector who had made some of the initial finds at Rouyn, of free gold near Woman Lake. Powell staked 9 claims there, and several other prospectors' syndicates staked a number of claims at Birch Lake, to the north-east. While Powell's discovery was initially publicized as being "in the Red Lake district" (as indeed it was, particularly from the viewpoint of the distant press), it was the beginning of a separate rush to that new area, which took much of the money, attention and manpower away from Red Lake.

A third faint negative note, one that went virtually unheard in the general optimism, came from Harry Darling. Darling was a mining engineer from Timmins who had left Hudson for Red Lake on March 20, with two men and eleven sleigh dogs, to stake claims. Darling returned to Hudson on April 11, and gave several interviews to newspaper reporters. In his conversation with the *Star* reporter, he sounded a prophetic note: "The snow staking of claims [at Red Lake] . . . may retard the development [of the area]. Many of the staked claims on snow will lapse, as no work will be done on them, and it will be a year before anyone else can reclaim."

May 1926

The waiting continued well into May–waiting for the snow to disappear so that prospectors could see what lay underneath, waiting for the lake ice to break up so that traffic could resume, waiting to see what the results of Dome's diamond-drilling would be. The *Toronto Star* proclaimed on May 8, "Fresh Stampede to Red Lake Will Start After Break-Up,", and went on to say that machinery, freight and men were all ready and waiting to go at Hudson.

And after the ice left, there was a surge of traffic out of Hudson again, this time on the water trail to Red Lake. The railway cars full of yowling sleigh dogs were replaced by cars full of canoes and boats, as dozens of vessels and their owners arrived every day. The CNR quickly built a large freight shed at Hudson to hold goods which were on their way to Red Lake. On May 12 Lac Seul was reported to be free of ice, and on that day, five prospecting parties left Hudson in the Hudson Bay Company's new boat, the Hulacapi.

The Hulacapi was one of three tugs, bought by the Hudson's Bay Company during the winter as the company reorganized its transportation business. On the Hulacapi's maiden voyage, it pulled behind it six York boats, including one transporting 75 Indian labourers who would be used to carry freight between Pine Ridge and Red Lake. A total of about 160 men (and one woman) were passengers in the flotilla, which was, in turn, followed by a number of canoes destined for use on the Chukuni. There were no problems on the trip, except for minor ones caused by the presence of the woman (who is, unfortunately, nameless). None of the boats had any sanitary facilities and the men, who were accustomed to simply drifting to the back of the boat to relieve themselves, were apparently embarrassed to do so on this voyage. (The woman's reactions went unrecorded.)

In that same week the Prospector–the Lac Seul Transport Company's new 40-foot launch–carrying 20 passengers, was launched, complete with champagne. The Triangle Fish Company's new 48′ launch, which could carry 20 passengers and 15 tons of freight, and was appropriately named The Triangle, was sent into the icy spring waters of Lac Seul for the first time. A new company called the Orser Red Lake line started operating smaller boats between Hudson and Red Lake.

The journey between Hudson and Red Lake by boat cost about $30, one-way, depending on the company; for example, the trip on the Prospector from Hudson to Pine Ridge cost $15, and a ride on the smaller boats of the Hudson's Bay Company from Pine Ridge to Red Lake cost another $15.

Jack Elliott, too, was back in business. The Elliott-Fairchild Company had purchased a motor cruiser, the Triton, to traverse Lac Seul. This was one of the largest boats on the lake, with sleeping accommodation for 15 passengers and space for another 35 on deck. It was in the Triton that the full complement of Dome directors journeyed to Red Lake in May to inspect the Howey property.

The Triton's first run on the Red Lake route was documented in detail by the press. The *Toronto Star* on May 22 carried a story, datelined Pine Ridge, describing the voyage. "With 8 canoes on her decks . . . and 32 people on board, the Elliott-Fairchild Company's motor cruiser Triton docked today after her first run on English River waters. She had left Hudson at noon yesterday. . . . Aboard were 31 men and one woman. The woman, Red Lake's first woman prospector, is Mrs. Mabelle D. Mars of Duluth. Clad in plus fours and a khaki overcoat, she carried an axe and moil [a piece of prospecting equipment] in her outfit."

The article goes on to say that Mrs. Mars was, in fact, a northern girl by upbringing–from Burk's Falls, near North Bay–and that she had an invalid husband, and no experience in prospecting. (Her story is a fascinating one, and more details of it follow.) The report also contains a detailed description of the

120-mile journey from Hudson to Pine Ridge, which the Triton traversed in about 15 hours travelling time, with an overnight stop at Lac Seul Factory, the Hudson's Bay post.

After the Triton's Red Lake-bound passengers disembarked at Pine Ridge, they were taken by the company's recently purchased Curtiss Flying Boat, piloted by Harold Farrington, to Red Lake. This aircraft, equipped with pontoons for landings and takeoffs on water, was the same type as those chartered by Jack Hammell in the previous fall to bring the first crew to the Howey property.

The Elliott-Fairchild aircraft was not the only one in the skies over Red Lake at this time. The Patricia Airways' Lark, piloted by "Doc" Oaks and Dale Atkinson, made flights to both Red Lake and Woman Lake. The plane, however, began to suffer from constant use and a lack of available spare parts, and its rather dangerous life was finally brought to an end when it did a complete somersault, on landing near its base at Pine Ridge. In addition, a new company, Central Canada Airlines, inaugurated a daily (weather permitting, as always) service between Kenora and Red Lake.

The first fatalities of the Red Lake gold rush occurred, not in the very cold northern air of winter, but in the less obviously dangerous icy waters of spring. At Hudson, hundreds of prospectors had been waiting for weeks for Lac Seul to be free of ice, and as soon as it cleared, they were ready to take off in anything that would float. On May 18, a scant six days after the ice was gone, a government inspector from the federal department of Inland Waterways arrived at Hudson to check every vessel to see that it complied with the minimum safety requirements.

Before he arrived, the task of inspecting the boats fell to OPP Constable Patrick Jago. Two young Englishmen, Stanley Quinn and Sydney Stebbings, had arrived in Hudson in early April to hit the winter trail to Red Lake. Fresh from a walking tour of France and Austria, they were eager to join in this wilderness experience in Canada and hastily assembled an outfit to go to Red Lake. Constable Jago got wind of their enterprise and, although they were certainly no less experienced than many who previously had mushed in to Red Lake, pronounced them "ill-equipped and utterly inexperienced". (Apparently they had failed to provide themselves with, among other things, any dog-food.) The Constable forbade them to go to Red Lake and, complying with his orders, they waited for open water.

While they were waiting, Quinn and Stebbings bought an 18′ canoe–and made a sail for it, which Jago burned. When open water arrived, they set off with the canoe well-loaded, and with Jago's admonitions to stay close to shore. Tragedy befell them in spite of these precautions, and when Neh Faulkenham arrived at Hudson on May 20–the first person to come out of Red Lake after break-up–he reported that their waterlogged canoe had been discovered floating on Lac Seul. The bodies of Quinn and Stebbings were found two weeks later.

While Hudson continued to blossom with commercial activity, there was increasing activity at Pine Ridge as well, and several more commercial and official enterprises established at Red Lake during the month of May. At Red Lake, a fellow from Sudbury opened a blacksmith's shop under a large poplar tree near Faulkenham's store. Bill Brown was appointed Red Lake's first postmaster, with his place of business, a tent, to be replaced by a log building (unkindly termed a log shack) which he built the following month. The Hudson's Bay Company decided to expand their operation at Red Lake, as well. They brought in two 560-pound "treaty tents" for a new post to be established opposite the Dome property on Red Lake. (These tents arrived as part of a mammoth shipment near the end of May, carried by 17 large canoes hitched together, with five outboard motors on the lead boats.) And a law office (in a tent) was established at Red Lake by two Toronto lawyers, Lawson and Treblecock.

Lawson and Treblecock also set up an office-tent at Pine Ridge, but that establishment shortly moved into a grander frame structure. Pine Ridge, like Hudson, was booming. From a trading post with six permanent residents, it had changed overnight into a bustling small town. There were 3 new stores, 3 restaurants, a hotel, 2 transport company offices, an Imperial Bank of Canada, a barbershop and pool room.

One of the "restaurants" at Pine Ridge specialized in the service of liquid rather than solid refreshments, and those of the alcoholic variety. There was also an illegal gambling operation there, and a business operated by two prostitutes, Goldie and Blanche. A great deal of money was spent in Pine Ridge, which came much closer to the popular idea of a free-wheeling gold rush town than the tent town at Red Lake did. Groups of claims changed hands frequently, with the buyer, usually a gullible newcomer, and the seller, a prospector passing through Pine Ridge, who had no trouble quickly spending the few dollars in his pocket.

The sudden development at Pine Ridge occurred because the village served, not just travellers going northwest to Red Lake, but those heading northeast to the Woman Lake area. The rapid increase in the number of claims being staked in the Woman Lake area led to another official development at Pine Ridge, and simultaneously signalled the end of the most intense phase of the gold rush at Red Lake.

At the end of April Charles McRea, the Ontario Minister of Mines, announced that, effective June 10, the Red Lake Mining Division would be enlarged to the east by adding on all the territory up to the boundary line between the districts of Kenora and Thunder Bay, and to the south as well. And on this date too, the Mining Recorder's office would be moved from Red Lake to Pine Ridge, to "thus meet the convenience of prospectors operating in the old district and a large part of the new."[40]

The actual move was completed before the end of May. Disgruntled prospectors at Red Lake once again had to travel some distance to file their claims, after the short-lived luxury of having a Recording Office on their doorstep. According to one of them, quoted in an article in the *Manitoba Free Press* on June 29, the distance from Woman Lake to Pine Ridge was only 6 or 7 miles shorter than that from Woman Lake to Red Lake, and the convenience of everyone concerned would have been much better met by leaving the office at Red Lake.

----

[40] *Manitoba Free Press*, April 28, 1926.

In the Woman Lake area, more free gold had been discovered, this time by two prospectors, named Clarke and Lester, on their claims south of Birch Lake. The first Woman Lake company was incorporated, Powell-Woman Lake Mines. The *Toronto Star* headlined on May 27, "WOMAN LAKE AND BIRCH LAKE GOALS OF VERITABLE STAMPEDE." The enthusiastic report said that 30 or 40 men, daily, were going up the Trout River (leading off from the Red Lake trail, north of Pine Ridge) and as many were going via the Wenesaga River (from Pine Ridge). The reporter included several anecdotes concerning the water trail to Birch and Woman Lakes. There were two Swedes who had no boat but instead built a log raft at the end of every portage and discarded it at the beginning of the next one; there was a well-equipped party of Americans, with huge canoes and a skiff, who carried tent-poles among their gear. (In that country there were tent-poles, albeit with leaves, every foot of the way.)

While much attention had shifted to the east, there was still a hum of mining activity around the quiet waters of Red Lake. Two more diamond drills arrived, to work on the Buffalo Red Lake and Cockeram properties. Gold was discovered (though in unknown quantities) on claims at East Bay, on the Victoria Syndicate claims, and on the Porcupine Goldfields claims. Major Cunningham-Dunlop reported that he had found free gold on the claims of one of his syndicates. Detailed maps of the Red Lake area, resulting from the government's aerial mapping of the area, became available to the eager public.

The Ontario Department of Mines sent in another party to do geological mapping in the area for the summer. In the words of the resulting report, "The discovery, in the vicinity of Red Lake, of gold deposits that give promise of commercial value and the great amount of prospecting that has followed these discoveries, have made necessary a re-examination of the geology of Red Lake . . . " The survey party, which arrived at Red Lake on May 22, was first headed by J. E. Hawley and then taken over by E. L. Bruce, who had done the initial geological mapping of the area in 1922 and 1923 that had led to Lorne and Ray Howey's gold discovery.

Diamond drilling by Dome was continuing on the Howey property with mixed results. At a depth of 500′, an unofficial report said that free gold was discovered in the diamond drill core, and stock units of the Howey syndicate gained a sensational \$50 in one day, on May 4. When the first official results were released, it was reported that one 66′ channel sample produced \$7.40 per ton in gold. (This grade, which was about .36 ounces per ton, might be described as a good average grade of ore, and in fact was appreciably above the grade of the ore eventually mined and milled by the Howey Mine.)

Provincial Geologist A. G. Burrows, like the stock market moguls, the mining companies, and the man in the tent at Red Lake, linked the future of Red Lake to the results of the diamond-drilling program. He said in a letter that "No-one yet knows how large the deposits in Red Lake are. The surface showings are exceptional. The [Dome officials'] commitments are so large that they must be satisfied that the results will justify them."[41]

The gold rush at Red Lake continued to be a source of interest to the American press. The Red Lake area was quite uncharacteristic of Canada, in 1926, which was already on its way, then, to becoming an urban nation. But it seemed to embody the American idea of Canada as a northern, untamed wilderness. (This might be called the Rose-Marie syndrome.) An example of this was found in a large, colourful wall map drawn by a Chicago artist for the annual convention of American Real Estate Boards, to be held that June in Tulsa, Oklahoma. On the map all roads in the U.S. led to Tulsa, exemplifying the importance of the upcoming event. And Canada was represented on this wall-sized effort by a musher, on the trail with his huskies, heading north, and a prospector sitting somewhere in the vicinity of Red Lake, saying "I hope they'll let my pipe through customs soon."

* * *

----

[41] *Toronto Star*, May 3, 1926.

The story of Mabelle D. Mars, who journeyed to Red Lake in May of 1926 on the Triton's first voyage north from Hudson, is an unparalleled one in the whole saga of the gold rush.[42] In April of 1926, Mabelle Mars was a stout, smiling and very enterprising woman of about 40, living in Duluth, Minnesota with her husband. Her husband, Charles Mars, was some years older than she. He was a Greek-born American businessman who had been one of the first manufacturers of candy and ice cream in Duluth, and he owned restaurants in Duluth and Superior. He had been forced to retire from his business some years earlier, as diabetes had caused, first a leg amputation, and then glaucoma. He still had a more than adequate income, though, as, according to Mabelle's diary in 1924, he received $1150.00 per month in "rent from business."

The presence of comfortable financial circumstances at home, along with her own early childhood spent at Burk's Falls, near North Bay, must have helped Mabelle decide that it was feasible for her to take part in the major North American adventure that spring, the gold rush to Red Lake. Before she actually journeyed to Red Lake, she had a dream about it. In this dream–the date of which she recorded, later, as April 15, 1926–it was revealed to her that there was a certain piece of land in the Red Lake area with a large amount of gold in the rocks beneath it, and that it was hers to claim and then develop as a producing gold mine.

She must have proceeded immediately to make arrangements to go to Red Lake to prospect. She initially planned to go with the prospecting party of an unnamed man from Port Arthur, but upon meeting him she realized he was not "the right type of man", and she decided to go on alone.[43]

----

[42] Much of the following account of Mabelle Mars has been pieced together from a diary of Mrs. Mars in the possession of the Ministry of Natural Resources at Red Lake.

[43] *Manitoba Free Press*, March 12, 1927.

As she told her story to the press, "I was to some extent familiar with open air life . . . since my hobby has been cross-country motoring. But I never had a taste of the real wilds before."[44] On her journey to Red Lake, she was the only woman on the Triton, and again the only woman on the second stage of the journey, from Pine Ridge to Red Lake. (On this journey she encountered the Reverend Maurice Sanderson, whom she described as a "remarkably intelligent man . . . with a fine tenor voice . . . [he] conducted daily services, praying and singing our own hymns translated into the Ojibwa language.")[45] Armed with a letter of introduction to Hudson's Bay officials, Mabelle "was the guest of the Hudson's Bay Company from Goldpines to Red Lake", she later told the *Free Press* reporter. (In the summer of 1926 the community of Pine Ridge changed its name to Goldpines.)

"Once I arrived at Red Lake", she recounted to the *Times-Journal* reporter, "I built a canvas tent, with log walls up to a certain height, and a tent over them, with a big bunk along one wall, where I spread my boughs. . . . I had a door to the tent, and I think that the lock I placed on it was the only lock at Red Lake."

Mabelle was undaunted by being the only woman–and without the protection of a husband–in a mining camp of men. As she said to the *Times-Journal* reporter, "How was I used by the people in the north? Why, they were all big brothers to me. . . . I should say that there is an opportunity for women in this mining district, if they know how to look after themselves." And she enjoyed the experience thoroughly. "Adventure? Well, the whole thing was an adventure. But there are some experiences that stick in my mind . . . One was the time we were lost all night through failing to recognize a portage and I and my men shivered on a rock from 11 at night until 5 in the morning around an open camp fire."

----

[44] *Fort William Daily Times-Journal,* August 6, 1926.

[45] *Manitoba Free Press,* March 12, 1927.

To the *Manitoba Free Press* journalist, Mabelle explained how she went about turning herself into a prospector. "To find my gold", she said, "I learned rock formations and how to recognize the different minerals . . . next I hired three men, canoes, Evinrude motor, tents, tools, and bought a supply of groceries."

After prospecting for three weeks, one day when she was exploring with her crew in the Post Narrows area near the north-east end of Red Lake, she apparently recognized in the rock-and-bush-encrusted surface of a small island, the rock formation she had dreamed about. The date was June 14, 1926, and she staked the claim immediately. The claim was numbered 4240, and on it were pinned most of this clever and industrious woman's hopes. She staked another two claims at Slate Bay of Red Lake, went out to the Mining Recorder's office at Pine Ridge to record them, then flew to the Woman Lake area to stake six more claims. She flew back to Pine Ridge to record these, and then home to Duluth.

When she was in Pine Ridge recording her second lot of claims, she took an unusual step. Unlike the vast majority of the participants in the gold rush to Red Lake, Mabelle Mars was aware of the fact that she was participating in a significant historical event. She therefore typed up and had six people sign, copies of an affidavit to the effect that:

> I ________________of Red Lake, District of Patricia in the Province of Ontario, make oath and say:
>
> 1. That I am personally acquainted with Mabelle Duncan Mars, of Duluth, Minn., in the United States of America.
> 2. That I have been a resident of the District known as the Red Lake Mining Division since ________________.
> 3. That to the best of my knowledge and belief the said Mabelle Duncan Mars, is the first white woman to enter this territory, and the first woman to record a claim in the said Mining Division and also the District of Patricia.

Two of the signees she enlisted were Bert Holland, the Mining Recorder, and J. W. McNeely. And it must be pointed out that Mabelle Mars was, as the affidavit stated, the first woman to record a claim in this mining division. However, she was not the first white woman in the territory. There had been at least one before her, the Mrs. Palo who had cooked for the Dome camp over the winter.

The next event in the story, as it can be pieced together, occurred in 1928. In February of that year, a Duluth artist, L. P. St. Germain, drew for Mabelle a diagrammatic X-ray map of her initial claim at Post Narrows, 4240. The map, apparently, is a concrete embodiment of Mabelle's dream two years earlier. This map, entitled a "Seance Drawing", indicates in detail where the ore-bodies (both gold and copper) on the property are located, where the main shaft should be dug, where drilling should occur, and where roads should be constructed. Handwritten notes on the map also indicate how many feet of water cover part of the gold pocket (the island is partly submerged), how deep the shaft should be, and the path of the veins (e.g. a note reads, "Vein 5 comes from Northwest, circles around 7 claims and returns to Northwest.")

In the summer of 1928, according to several entries in one diary, Mrs. Mars forwarded a copy of this map to Senator Archie M. Chisholm, a friend of hers. According to her notes, Senator Chisholm then forwarded his copy of the map–along with the hefty sum of $15,000 of his own funds–to his cousin who was then at Red Lake. The money was to be used for preliminary development work on the property, using the map as a guide.

In 1929, Mabelle made two trips to Toronto in her attempts to get help in developing the property. She also returned to Red Lake in July of that year. At this point she staked three more claims in the area of 4240, the Post Narrows or "Spirit Map" claim, as she referred to it. She also discovered, according to her notes, that no work had been done on her property with the money sent by Archie Chisholm for the purpose. She had some work done on the property, herself, while she was there.

An article on her claims appeared in the *Toronto Mail and Empire* on October 4 of that year. "Mrs. Mabelle Mars . . . has recently returned from her property, which is known as the Mars Basin mine. The location consists of 4 claims and takes in Mars Island, which is located approximately 7 miles north of the Howey Mine. . . . The property has already been proved up to some extent and trenching and stripping carried out. . . . While no diamond drilling work has as yet been undertaken, Mrs. Mars is strongly of the opinion that she has a real mine in the making."

In that month, the well-known stock-market crash occurred. In Mabelle's words, "all Bay Street said Red Lake must wait until all this crash is built anew and that gold at $20.60 an ounce could never be mined in Patricia . . . at a profit . . . and when I returned home my husband was totally blind, I had to give him his insulin shots and strap on his leg. My mining adventure was over."

This particular entry seems to have been written many years later (the last entry in this particular diary, in fact, dates from 1958) and the note of resignation it contains occurs rarely. Mabelle obviously wanted to return to Red Lake herself, and she was torn between her love for the free-wheeling life in the north, with its chance for her to develop her property on Mars Island, and her duties to her invalid husband. Another entry says, "I asked God what I should do. He said my husband was my first obligation, that I should take care of him, this I have done very well."

There are no references to any further trips by Mabelle to Red Lake until 1949, by which time all her properties had long since reverted to the Crown and been restaked by others. She did not cease, though, in her efforts to have what she called the "Spirit Mine" developed. Duty-bound to her invalid husband as she was, she notes in her diary that she wrote to Judge Godson of the Ontario Court of Mines to see what she should do to ensure that her staked properties remained in her name. The judge replied, she notes, that "he would personally see to it that my properties were protected as long as my husband was an invalid." She also tried a co-operative approach. "I asked the Duluth Chamber of Commerce to try organizing a syndicate of their members to take up where I left off and use a percentage of the earnings every year for the Duluth Community Chest, but nothing was done about it as far as I ever learned."

Judge Godson's assurances to the contrary, her properties did not remain protected and, according to her notes, the 4240 claim was restaked for himself and J. R. Brodie, in 1940, by C. D. Chisholm. (She said he had, in the interim, been sent another large personal loan–this one for $40,000–by his cousin Senator Archie Chisholm, Mabelle's friend.) This property, along with other neighbouring ones, for a total of 18 claims, was known as the McFindley Mine and financed by Little Long Lac, the diary says, to the tune of $4 million. It produced virtually nothing–because, says Mabelle in her notes, the operators "had not known [that the gold-bearing vein of the property] . . . was a circular job", as shown in her spirit map.

Mabelle obviously had frequent contact with the promoter John E. Hammell, for she implies, in several places, that he told her mining was no business for a woman, that she was "in the doghouse with John E." She mentions in an entry from the late 1940s, with an understandable note of satisfaction, that the "South Shore [mines–i. e. Hammell's Howey and Hasaga properties] folded up. Just as I said they would. Red Lake is a little world of little people."

It is not possible to round off Mabelle Mars' story with a happy end that tells of 4240 turning into a major producing gold mine. On the other hand, her spirit map still exists, should anyone want to consult it. And as for Mabelle, she was a woman of strong determination and energy who did secure for herself a place in the history of the gold rush to Red Lake. Perhaps a sentence in her diary typifies her character: "Failure is not defeat, it is only getting experience."

The Gold Rush, Part V: June-December, 1926

As the month of June began, Red Lake hummed with activity. The big companies were very active. The three steam-driven diamond drills at the Howey property were working 12-hour shifts around the clock, and there were two more of the noisy metal monsters at the McIntyre-Porcupine property, as well as one each at the Buffalo and Cockeram properties. It has been estimated that Dome, McIntyre, Dunlop and Huronian Mines together employed about 400 men at this point. Hundreds of others were working in smaller groups of 8 or 10 on their claims–for example, Noranda Mines sent 10 men to work on its group of claims; Bill Tyrrell returned with 8 men to start development work on his Slate Bay property; the Red Lake Exploration Company had a crew of that size on its property about a mile north-east of the Howey, and 10 men left Toronto for Red Lake to work on the Miller-Glick claims.

Many of the one-man operations, though, were simply putting in time, as they waited for the results of Dome's diamond drilling to increase the value of their claims. This was not the case for all of them, however. Because the moss on the ground on the heavily-wooded areas was still frozen solid, prospectors who wanted to work their ground on such claims were lighting fires to thaw out the ground before they attempted, with pick and shovel, to find out what lay underneath.

The stock units of the Howey Syndicate varied considerably in price from day to day–from $185 on June 11 to $255 four days later, for example. The general feeling about the future of the Howey, and of Red Lake, was still positive. On June 15, the *Star* reported that Doug Wright (chief geologist at Dome in charge of Red Lake operations) had called on the Minister of Mines and told McRea that he was just as optimistic about Red Lake as he had been at the beginning. At the end of June it was reported that Bill Smith (the manager of the Hudson's Bay Company post at Red Lake) had found gold on his claims, east of McKenzie Island and north of the Cochenour property. Burrows, chief geologist of the Ontario Department of Mines, had gone to Red Lake in June for a tour of inspection, and said that this further new discovery had convinced him that Red Lake had an important future. Burrows remarked that the new discovery was just what was needed to revive the enthusiasm of some of the miners who had run out of both money and faith, and were letting the assessment work on their claims go undone.

During the month of June Red Lake was continuing to acquire, however slowly, some of the trappings of civilization. Dominion land surveyors were at work and laid out seven townships, each six miles square, in the area. Bill Brown, the postmaster, moved from his tent office into one built of logs, with a tent beside to serve as living quarters. He later added a small extension for a general store and restaurant, and then another for poker games.

It was also at this time that Red Lake came to acquire its first manufacturer of what was called "home brew". Apparently this entrepreneur had set up his tiny tent and equipment on a small island out in Howey Bay. His establishment was so situated that, after rolling up the walls of his tent, he could lie in comfort on his bed roll with a clear view of all approaches to his island. And if any official-looking canoe came his way, he could get rid of the evidence by kicking his equipment straight into the lake. (This anecdote is typical of the light-hearted approach to the Prohibition-induced manufacture and sale of illegal alcohol at Red Lake. Much less light-hearted, though, were the frequent cases reported in the newspapers at this time of unwitting drinkers of "home brew" who had been blinded or even killed by drinking the stuff.)

Another civilizing factor–at the opposite end of the spectrum from the poker/home brew scene–was the first church service in Red Lake. This was conducted by Canon Maurice Sanderson, a minister of the Anglican Church who was then stationed at the Anglican mission at Lac Seul. Canon Sanderson, who was a full-blooded Indian, had graduated from St. John's College in Winnipeg and been ordained a priest in 1902. He had a long and very honourable history of working with and ministering to both the Indian and white people scattered across northwestern Ontario, travelling by canoe in summer and by dogsled in winter to call on his parishioners. On a Sunday morning in June, Reverend Sanderson held a service under the trees in front of Bill Brown's new post office, and over a hundred men gathered to participate in the ecumenical service.

It is impossible to know exactly how many men were going to Red Lake during the month of June. We know that the numbers were well down from March, but were greater than in late April and early May, when spring break-up hindered all travel. The traffic to Red Lake in June was at least sufficient for the Department of Natural Resources of the CNR to issue a statement, quoted in the *Canadian Mining Journal* of June 11, responding to the hundreds of inquiries from across North America that it had received.

The purpose of this statement, said the Department, was to give "general ideas of the conditions prevailing and of the means of access" to the Red Lake area. The statement said that ample provision was being made for the transport of passengers and freight by gasoline launches between Hudson or Sioux Lookout and Pine Ridge, and that at Hudson freight could be loaded directly from the railway siding into boats. There was boat service from Pine Ridge to Red Lake, the statement said, and other alternatives too: by plane all the way, by a combination of boat and plane (with the Elliott-Fairchild service), or by canoe alone, either from Hudson or via an alternative route that left the railway line about 80 miles west of Hudson, at Canyon Lake.

The number of would-be prospectors going to Woman Lake was not as great as that still heading to Red Lake, but nevertheless Woman Lake received substantial publicity. The *Toronto Star* on June 5 carried a front-page story headlined "GOLD STRIKE AT WOMAN LAKE MAY SURPASS RED LAKE FIND", and explaining that some samples from there reportedly showed values of an incredible $72.50 per ton. A few days later, on June 8, there were reports of claims being recorded from "the first big strike made at Birch Lake", with "results from panning reported spectacular." Free gold was reportedly discovered on the Rowen claims, between Woman and Clearwater Lakes. The gold finds at Red Lake and Woman Lake had caused a general revival of interest in the staking of mineral claims in northwestern Ontario, and the Mining Recorder at Kenora reported that there were now over 30 claims daily being recorded there.

There continued to be optimistic talk about a more straightforward method of transportation to the new northern gold fields. A party of land surveyors began to survey the route for a road north from Quibell to Red Lake. And Bill 95, for the construction of the Red Lake and North Western Railway, was again debated in Parliament. The point was made that the responsibility for building such a railway belonged not with the federal government but with either the CNR or the Province of Ontario, and the bill was "talked out" without coming to a vote, though it remained on the order paper.

The most unusual assortment of passengers and freight to travel to Red Lake by any method, at that time, was the Red Cross party. The medical facilities at Red Lake then consisted of a doctor, W. S. Fitzpatrick, employed by the Dome and McIntyre Mines to treat those companies' employees. The nearest medical service for anyone else (although presumably in cases of emergency the Dome doctor could be called upon) was the Red Cross car at Hudson.

OPP Constable Carr was one of the people who made specific requests to the Red Cross to extend the outpost service to Red Lake, saying, in his letter of request, that there were no medical services within a hundred miles of Red Lake. (When Carr wrote his letter, it was before the arrival of Dome's first resident doctor, in mid-April.) Carr also voiced his concern about the lack of medical facilities at Red Lake to the press, saying that the primitive methods of sanitation in use there could cause severe health problems when large crowds arrived.

The Red Cross moved speedily to establish the new service. After the decision to establish an outpost hospital at Red Lake was made in April, Dr. Fred Routley, the Ontario director of the Red Cross, flew to Red Lake at the end of that month. He spent several days there to determine just what medical services were necessary, and what facilities would be available for a resident medical staff. (The second point would be easy to determine, the answer being, virtually nothing.) According to the press, arrangements for shipping all the equipment were made by Clara Wheatley, the nurse in charge of the Red Cross emergency outpost in the railway car at Hudson, and by Jessie Gibson, the other nurse at Hudson, and daughter of the Deputy Minister of Mines for Ontario.

Dr. Routley asked Dr. Dingwall of the Dryden Red Cross hospital to accompany the nurses to Red Lake. The party, as it assembled at Dryden before embarking for Red Lake, consisted of the two nurses who were to staff the Red Lake outpost, Agatha Gamble and Alma Finnie; an orderly named Papke; Maude Wilkinson, Director of Nursing for the Red Cross; and Dr. Dingwall and his wife. They drove from Dryden 12 miles north to the CN line, and then took the train east to Hudson. There they met with their supplies, spent the night at the Red Cross car, and set off the next morning, June 26, for Red Lake.

The departure of the Red Cross party for Red Lake was of great interest to the press, and in fact it was one of the most interesting episodes in the summer months of the gold rush. The supplies and equipment that had been assembled for the Red Lake outpost totalled some 7,000 pounds, and the small mountain of goods was packed into five large freight canoes owned by the Hudson's Bay Company. The canoes were manned by eight Indians, and towed by one of the Hudson's Bay Company pointer boats, equipped with an outboard motor and carrying the medical group. The first night they reached the Hudson's Bay post at Lac Seul, which looked like a miniature oasis of civilization, featuring neat white buildings and a small golf course. Frank Aldous, the Factor, and his wife entertained their guests, put them up for the night, and sent them on their way with some canned moose meat.

Because the waters of Lac Seul often became rough in the late afternoon, it was necessary to do the bulk of the day's travelling early, and the party set out the next morning before five o'clock. After travelling steadily all day, the party reached Ear Falls and was crossing the second portage there when they met a well-dressed gentleman, accompanied by two more typical prospectors, going in the other direction.

The gentleman in question was obviously surprised to see four women in city clothes on the trail, and told them they would never make it over the other portages to Red Lake, as the mud was hip-deep in several places. Dr. Dingwall stoutly replied that he would get them over if he had to carry them on his back. The gentleman then inquired why these ladies were going into a rough mining camp, and, when told that they were nurses who were going to open an outpost hospital at Red Lake, replied with his congratulations. It seems that he was a doctor from Toronto himself, who was on his way back home after inspecting his claims at Red Lake.

The rest of the journey continued without incident–the nurses photographed a bull moose at one point, and were able to circumnavigate the mud-holes on the portages without much difficulty. The party arrived in what is now Howey Bay of Red Lake on the afternoon of June 30, but were not quite sure where to land, as there were only two buildings and no docks at all in sight. (The two buildings were Bill Brown's post office and the Dome Mines Assay Office where the company doctor, Dr. Fitzpatrick, lived.) The party eventually landed, and managed to find Victor James, assistant manager of the Dome operation. (They were supposed to have been met by Constable Carr, who had offered to arrange accommodation for them, but he had been called away to Fort William.)

Nevertheless, this unusual group was rapidly made at home. James ordered a storage tent erected for the party's supplies. The four women moved into Dr. Fitzpatrick's quarters, in the assay office cabin, and Dr. Fitzpatrick, Dr. Dingwall and the orderly occupied Constable Carr's tent. Victor James had eight men clear a 100′ x 250′ lot for the outpost hospital, and Dr. Dingwall (accompanied by Papke) took axe and saw in hand and cut a trail to the post office. The new trail was named when someone nailed a piece of board to a tree with the words "Dingwall Avenue" on it. This trail, in fact, was Red Lake's first street, and later became the main street of the town, called Howey Street.

When the facilities had been set up, Dr. Dingwall and his wife and Maude Wilkinson flew out to Pine Ridge in the Lark and then rode in the Prospector launch to Hudson. Nurses Finnie and Gamble remained in charge of the hospital, the only women (with the occasional exception of Mabelle Mars) in a mining camp with a shifting population of a thousand men. Neither of the nurses were at all fazed by the situation, however, as–like those other pioneers of the gold rush, the bush pilots–both of them had served in World War I, as nurses in the Canadian Army Medical Corps.

The outpost hospital was maintained only for the summer, as it turned out, and during that time 13 patients were admitted for treatment, for a total of 93 hospital days. (There was an unrecorded number of patients treated but not admitted). One of the patients was a prospector with a severe case of arthritis who was also unhealthily fat, weighing some 240 pounds. When asked by Nurse Finnie how he managed to get around in the bush, he replied that he had come in not to stake or develop a property, but to look the camp over with a view to establishing a bootlegging operation–he discovered, though, that there was too much competition to make it worthwhile.

Able-bodied patients at the outpost hospital were expected to help out with the cooking and one young man, a novice at the art, decided to make rice pudding. The experience was like the overflowing porridge-pot of fairy-tale fame, as he filled the pot with rice and, when it began to boil, had to use every dish in sight to hold the cooked rice–ending with his own hat.

The nurses were well provided for, in terms of food, during their stay. The McIntyre camp often sent them fresh fish and bread from their cook. A man from the Dome camp brought them some venison. They kept their dried milk powder in cans dug into the ground and banked over with moss, and mixed a daily supply with fresh water. Jack Hammell once brought them a box of fresh tomatoes from the matron at the Dryden Hospital–their only fresh vegetables at Red Lake.

The nurses each had a dog, for companionship and for protection against wild animals. (They kept a pail with rock sugar near the door for bears who might come in the night.) The bears stayed away, but the nurses' own dogs were often attacked by packs of the sleigh dogs who had been left to fend for themselves by the prospectors who had brought them in (and paid very high prices for them too). The short northern nights were dark and quiet, lit only by the occasional oil-lamp or candle in one of the camps, or a prospector's camp-fire in the distance. It was an independent life, and Nurses Gamble and Finnie were among the very few women who experienced the pleasures of it.

The month of July, the month of high summer at Red Lake, was very much like the preceding month in the saga of the gold rush. At Woman Lake there were further finds which drew prospectors off from Red Lake to the east; at Red Lake there were further discoveries reported as well, but the general "wait-and-see" mood continued as diamond drilling proceeded.

Reports of new discoveries and new development work at Red Lake continued to circulate. A crew arrived at Red Lake on July 7 to do sampling work on the Hollinger Mines claims staked during the winter. On that date the *Toronto Star* reported an interview with a "veteran prospector" who predicted that the Red Lake area would have a population of 20,000 within ten years. The venerable Noah Timmins, of Timmins fame, optioned a group of claims which were reported to have showings of free gold. On July 19, a rich discovery of gold was reported on the Rivers Syndicate claims at Red Lake. On the 22nd, important gold finds were proclaimed both at the Victoria Syndicate claims on East Bay and on the McIntyre Claims, and on July 26, another find was reported in the Wolf Narrows region of Red Lake.

Such reports were sufficient to keep people and goods flowing in to Red Lake. Patricia Airways and Exploration, Captain Oaks' company, remained busy, and advertised in the Toronto papers its "Daily Service from Sioux Lookout to Pine Ridge, Red Lake, Woman Lake, Birch Lake and Cat Lake." (It also advertised its Aerial Mail: 25 cents a letter to Pine Ridge and Red Lake, 50 cents to Woman Lake and Birch Lake.) The planes continued to perform their emergency/mercy function: when a prospector named Harry Jones lost his way prospecting around Narrow Lake, Captain Oaks spotted his campfire two days later and flew him back to Red Lake.

The Red Lake Transport Company (the McDougall and Wardrope enterprise) put into operation a type of stage-boat service. They bought three new motor-driven barges, stationed one above each of the three long portages between Pine Ridge and

Red Lake, and also kept a four-horse team at each portage to haul freight across the rough portage roads. And–to move from the actual to the hypothetical, in terms of transportation–Bill 95 concerning the Red Lake railway came before the House of Commons Committee on Railways and Canals. The majority of the Committee's members were of the opinion that such a railway would become yet another white elephant and would have to be absorbed by the CNR. The Bill was defeated.

Because of the intense American press coverage the Red Lake rush had received, letters continued to pour in to the Mining Recorder for the area, now at Pine Ridge. One such was from a woman in Miami, Florida. She told the Recorder that she and her husband wanted to drive to Red Lake in their motor car. At Red Lake she would run a hotel or a restaurant while her husband, a semi-invalid, did "light work in the open air". Since they were both vegetarians and must have proper fresh vegetables, she added, would this pose a problem? (Indeed it would, considering that the Red Cross nurses' only fresh vegetables in three months consisted of a box of tomatoes.) Another letter was from a group of three New York City schoolteachers who wanted to know if the Recorder could find them some work, as they'd like to come to Red Lake for their summer holidays.

In the press of July, though, Woman Lake had precedence over Red Lake. The *Toronto Star* on July 5 announced "GOLD EAST OF WOMAN LAKE FOUND BY TORONTO MAN; Rush to New Goldfields Leaves Red Lake Quiet." The following day, it carried a report of an interview with a prospector who had dumped all his mineral samples from his claims near Woman Lake–with apparently high values of gold and silver–on the floor of his hotel room for the reporter's edification. Throughout the month the departures of parties for the Woman/Narrow/Birch Lakes area were frequently covered by the press: "NEW GOLD RUSH STARTS TOWARDS NARROW LAKE FIND" and "NARROW LAKE FIND IS PATRICIA'S GREATEST" were typical headlines.

The enthusiastic reports of discoveries at Narrow Lake remained both unquestioned and largely unsubstantiated: the rush to that area was at the same stage of development as the one to Red Lake had been some six or seven months previous. But at Red Lake, by now, eager reports were not enough. The area had already been almost completely staked (with the overwhelming majority of the claims on worthless land.) It remained to be seen just what the area was actually capable of producing. As an article in the *Toronto Star* said on July 5, "the rush to Red Lake has settled down to a waiting game which the action of the Dome Company might . . . end. Stakers by the hundreds have failed to . . . do the assessment work on their claims which must be done by September 21. . . . Little actual work is being done at Red Lake, though some companies are bringing in men to do the assessment work."

All eyes were on Dome's activities, but none were able to predict what Dome's decision would be. After a short break, diamond-drilling on the Howey property resumed on July 13. Only the financially naive, though (and there were plenty of these), believed that the results of the diamond-drilling would be the only factor in Dome's decision to take up or drop its option at Red Lake. The stock of Dome began to drop rapidly around the beginning of July. The stock, which had sold at $20.19 in February, dropped to a low of $10.25 on July 13.

There were a number of reasons for this: the ore recently mined at the Dome Mines property in Porcupine had been reported to be of inferior grade, and–for this and other reasons–there had recently been large numbers of the shares offered for sale. Dome's involvement at the Howey had also prompted the sale of many shares; as one broker remarked to the *Star* on July 16, "Some people sold because they were afraid Dome would not take . . . [the Howey option] up and others because they were afraid they would . . . [the people who sold stocks for the latter reason were more concerned about] the cash outlays required of Dome than the possible profit-making asset it would be if Howey turned out well." Strenuous efforts, apparently, were being made to keep the stock price above $10.00, for, if it fell below that, it wouldn't be considered good loan collateral on Wall Street.

In addition to the cautions about the Red Lake situation which were caused by Dome, there were others of a more general nature being voiced. Unnamed geologists were quoted in the *Toronto Star*, on July 9, as saying that there was no doubt that there would eventually be several producing mines at Red Lake–but that this development might happen after the boom had subsided.

Doug Wright, Chief Exploration Engineer for Dome and in charge of work at the Howey property, voiced the same idea at much greater length in an article he wrote for the Engineering and Mining Journal Press of New York.[46] In general, he said, the area looked promising, but it was necessary to wait and see. Only one-tenth of the 3000-odd stakings at Red Lake, he said, were not snow-stakings (that is, claims staked completely blind, with no idea of the mineral formations involved). "Red Lake," said Wright, "has been heralded as El Dorado found again, and the daily press has gratuitously super-advertised the area. . . . Such a situation is far from the facts."

Wright did not imply that the Red Lake area lacked mineral potential, but rather that its wealth was of the kind that takes many years to bring to fruition. He cited a number of mining areas that had been discovered, had "boomed", but then had taken years to really develop. "Rice Lake, discovered and boomed in 1912, is in 1926 starting to respond to intelligent development; Gowganda, discovered and boomed in 1908, has . . . just come into its own . . . Kirkland Lake, Ontario's second most important gold-producing area, was discovered in 1912 but only in the last 4 or 5 years has its importance been appreciated . . . camp after camp has had to go through the vicissitudes of 'booming'. Is Red Lake due to suffer in the same way?"

Subsequent events proved Wright to be precisely correct. In the month following the publication of his article, the Red Lake boom collapsed. And, although the Howey mine did come into production just four years later, it was close to 15 years after the Howey brothers' initial discovery before the Red Lake area could be called a major gold producer.

----

[46] Quoted in the *Toronto Star*, July 2, 1926.

It was in the month of August that what seemed to be a death knell sounded for Red Lake. And it was also in August that the sudden death occurred of one of the rush's major figures, Nehemiah Faulkenham. One of the standard commodities stocked by Faulkenham's store was dynamite, a necessity for claim-holders attempting to do development work on their properties. In early August, Faulkenham came into possession of a case of defective sticks of dynamite. This very dubious commodity had apparently been in contact with water at some point on its trip to Red Lake, as some of the nitroglycerine had leached out and recrystallized on the paper coating of the sticks. For some days the case lay on the ground just outside the store; Faulkenham is reported to have told someone, at this point, that he knew he should move it away but, to tell the truth, he was afraid of it.

He decided he would cache it on the shore of the lake, 200′ away from his store, and maybe use it for his own claims. When last seen alive on August 10, he was carrying the box. In the explosion that followed, apparently just two of the sticks ignited, but Faulkenham was killed instantly. A customer, who was loading his canoe, ran back at the explosion and found the body. An inquest ruled that Faulkenham had died an accidental death and his body was taken by train to Matheson where he was buried. Faulkenham's recently acquired partner, Frank Stopps, closed the business and sold all the equipment and supplies Faulkenham had accumulated, including all the lumber he had carried in to Red Lake, to Gus Dupont, who took them by boat to the claims he was developing at Pipestone Bay.

Faulkenham's death was a real blow to the rudimentary community at Red Lake. The wiry little man, one of the real pioneers at Red Lake and the man who knew the trail better than anyone, was known for his generosity, both before and after he opened his store. (The *Toronto Star's* August 13 article on his death quoted a Red Lake prospector to the effect that Neh never turned a man down who didn't have the money to pay for his goods–and there were a lot of those at Red Lake that summer.)

The press reported Faulkenham's death; they reported, in early August, that there were forest fires on the trail to Red Lake (including one with a one-mile front burning on Pakwash Lake, another with a two-mile front burning at Keg Lake); and they reported, throughout August, a number of new gold discoveries in the Woman/Narrow Lakes area (REPORT NEW STRIKE IN NARROW LAKE . . . SAYS WOMAN LAKE FIND AN IMPORTANT ONE . . . MORE ENCOURAGEMENT FROM WOMAN LAKE). The really big news from Red Lake, though, was buried on the inside pages–this was Dome's decision to drop the Howey option.

On August 12 the *Toronto Star* carried a small article, in its Mining Stocks section, to the effect that Dome Mines would probably continue its present high dividend rate of $2.00 per share for another quarter, even though the earnings at the Dome property in Porcupine were down. The article stopped there, but the conclusion might well have been drawn that the corporate decision to keep dividends high did not make exploration possibilities–like the Howey property–very likely.

On August 16, the news broke, though there was no official announcement: Dome would not be taking up its option to purchase the Howey Syndicate claims. (And on the breaking of the news there was a tremendous amount of trading in Dome shares, as well as those of the other major mining companies, with Dome's shares dropping 40 points.)

On the evening of August 17, though no official announcement had yet come from Dome, Jack Hammell handed out a statement to the press to the effect that the deal was off. The *Star's* reporter came up with background figures showing that Dome had spent about $150,000 on the Howey property before it decided not to go ahead with it: $25,000 on preliminary work, $75,000 on cash payments for the right to the option, and another $50,000 on drilling and other exploratory expenses.

The indomitable Jack Hammell was cheerful as he spoke to the group of reporters. "No statement can be made at the moment," he said, "We have several negotiations on tap. The engineers are going over the data and compiling figures. . . . Look me over and see if you think I look downhearted. There are lots of things we can do. We might work the mine ourselves. . . . We've got a mine, on Dome's figures."

On the 19th, the *Northern Miner* carried Dome's official statement that it was dropping its Howey Red Lake option. This statement read, in part, that "although Dome . . . has secured interesting indications . . . in Red Lake, these have not been extensive and important enough to warrant belief that the expenditure involved will be repaid. Dome therefore is dropping all its options and properties in Red Lake. These include Howey and Red Lake Prospectors. The claims staked by Dome will be retained."

On the same day the fortunes of Red Lake received another setback. As the *Toronto Star* of August 19 put it, "J. P. Bicknell, president of McIntyre-Porcupine Mines, in the annual report mailed to stockholders today, hits another blow at the Red Lake camp." The article quotes a section of the annual report: "Last summer . . . the company acquired a 90% interest in a group of claims situated in the Red Lake district.[These were the claims staked by Ray Howey.] After examination our engineers recommended a program of exploration but so far the results of the preliminary work have failed to confirm our expectations."

Two more developments followed immediately on Dome's action. The first was a formal matter. At a meeting of the stockholders of the Howey Syndicate, the representatives from Dome resigned from the Howey board. The second development–and the only remotely positive one–was that Doug Wright resigned from his job with Dome because they had dropped the Howey option. Wright believed that the Howey property was sufficiently rich to make a profitable gold mine, and announced that he would be linking his fortunes, in that belief, to those of Jack Hammell and the Howey Syndicate.

The effect at Red Lake of Dome's dropping the Howey option was drastic and immediate. If a wealthy corporation like Dome didn't think a profitable mine could be made out of a property with good gold values like the Howey, then why should anyone else bother hanging around Red Lake? The exodus of prospectors from Red Lake was swift. Many went over to Woman Lake (if they hadn't already been there) to stake some claims before freeze-up; others simply loaded up and headed back down the trail to Hudson.

There were a few last flurries of excitement before winter and silence descended again upon Red Lake. A significant discovery of gold was reported at the end of August on a group of claims at East Bay. These claims were to come open for restaking at midnight on August 31, and a die-hard group of over 40 men waited there for the clock to strike midnight to begin staking. New gold discoveries were made in September at the McManus Red Lake claims, on the Chukuni River. Part of the crew at Dome had been moved to Dome's own claims, about three miles west of the Howey group, and several finds of free gold were reported there.

The reports of the new Dome discoveries were greeted with some cynicism–even the *Northern Miner* reported the then-circulating rumour that the Dome management had known about the good values on their own claims for some time, and had been influenced by these to drop their Howey option. To other of the old-time prospectors, this confirmed their anti-corporate suspicions: after Dome dropped their Howey option and most of the other claims were consequently left unworked and up for restaking, they reasoned, Dome would then buy up many other properties for a song. There was, however, one recorded note of optimism. The invincible Major Cunningham-Dunlop was quoted, in the *Toronto Star* on September 1, as saying that the new find on the Dome claims was "the richest made in the district" and should lead to a revival of interest in Red Lake.

All such excitement, though, was short-lived and did nothing, ultimately, to prevent the sorry exodus of prospectors from Red Lake. Dome disbanded its operations and left in a flotilla of boats and canoes in late September. The Red Cross outpost stations at Red Lake and Hudson were closed at the end of September. The Hudson's Bay Company post, which had recently moved, remained open but its lowered prices reflected the drastically lowered demand for its provisions (as well as the lower costs of summer freighting). Butter that had cost \$1.50 per pound in March was now \$.75 per pound; tea had dropped from

$1.35 to $.75 per pound; the price of bacon had fallen from $1.25 to $.50 per pound.

Gold discoveries in the Woman/Narrow Lakes area continued to draw press attention and prospectors in the few remaining weeks before freeze-up. A prospector named Tom Bathurst made a spectacular find near Narrow Lake. While cutting his claim lines, he came upon a large outcropping of quartz veined with free gold. One piece of this gold, reportedly, was one and a half inches long and half an inch wide. Bathurst Gold Mines was rapidly financed and incorporated. At Woman Lake, it was much like the scene at Red Lake in the spring–but quite unlike the scene at Red Lake in the fall.

The water-route from Red Lake to Hudson was thick with boats eager to hit Hudson before freeze-up. This armada in retreat included, though, a few men who wanted to pick up supplies at Hudson and return to Red Lake for the winter. There were a few prospectors who were going to stay and work their own claims; a few still employed at some of the larger camps, like the Dunlop one, cutting wood and generally preparing for development work the next spring; and a few who decided that they might as well stay at Red Lake and trap for the winter since they didn't have a job anywhere else. Gordon Shearn was one of the employed, and Conrad Hanson, one of the trappers, who elected to stay. According to Hanson, there were 26 people left at Red Lake directly after freeze-up.

Mail service dwindled to a delivery every ten days, brought from Hudson by dog team. On Saturday nights, the isolated men used to congregate at the Hudson's Bay post for the evening. As the winter set in, the small number of inhabitants at Red Lake became even smaller. Christmas, 1926 was celebrated together by the group of about ten men who remained at Red Lake. They gathered at the Hudson's Bay post for the occasion. By some standards, they were a sorry group–the flotsam left behind on the beach when the tide of the gold rush had come and gone. But some, like Gordon Shearn, were still young and enthusiastic, and others, like Sandy McIntyre, who had seen several major prospecting discoveries slip through their fingers, with typical optimism, were ready for the next one to turn up.

Apparently someone brought several gallons of home-made beer to the Christmas Day celebration at the post, and Christmas cheer prevailed. A rifle meet was held in the afternoon, and the small group dispersed to their cabins before dark. It was Christmas Day of 1926, and the Red Lake gold rush was over.

After the Gold Rush

The immediate result of the gold rush–of that mass movement of thousands of men, and the expenditure of a very large sum of money–was virtually non-existent. The Ontario Department of Mines geologists who spent the summer of 1926 in Red Lake, described the situation in their report published in the following year.

"Reports of [the Howey brothers' gold discovery] led to a rush of prospectors . . . and to the staking of most of the surrounding country. . . . During the summer of 1926, some development work and intensive prospecting was done . . . but the discoveries reported from the Woman, Birch and Narrow lakes areas . . . diverted the attention of prospectors from Red Lake. There was a tendency on the part of many claim-holders to wait for the results on the Howey and McIntyre properties before undertaking expenditures on their own claims. In many cases . . . individuals had staked claims in the hope of a quick sale and found themselves burdened with a large area, which they were quite unable to prospect properly. Some prospectors, and especially many of the mining and prospecting companies, had their claims surveyed in the latter part of the summer, satisfying in that way the requirements for the first year's assessment work. As a result, the amount of actual work . . . was not great."

In this respect the Howey property stood alone, for it was the only property in the area to have significant work done on it. And the Howey was to stand alone for most of its producing life as well–as the first producing mine in the Red Lake area, and hence as the one indirectly responsible for the ultimate development of all the others.

If Jack Hammell had not persistently set about to refinance development of the Howey property, there is little doubt that the Howey would have gone the way of other properties that had been optioned and then dropped by major companies (and that route led straight to nowhere, in terms of development). Hammell, though, had financed the mammoth project at Flin Flon, and was not to be deterred by the smaller Howey project.

He reconstituted the Board of Directors of Howey Gold Mines Limited. He was President; R. T. Birks the Secretary-Treasurer (and later President), and also on the Board were Harry Oakes (later Sir Harry, staker of the great Lake Shore mine at Kirkland Lake), Al Wende (founder of the Wright-Hargreaves mine) and W. S. Cherry.

Cherry proved to be the financial backbone of the enterprise, both in its newly reborn state and later on in its development. Cherry was a Canadian-born merchant magnate from Rhode Island who initially agreed to pay $100,000 for 1/4 of the 1 million Howey shares. Horace G. Young, a mining engineer who later had his own short-lived gold mine at Red Lake, was hired as general manager of the Howey, Doug Wright as chief geologist, and Hammell was ready to go ahead by February of 1927.

At that point the almost-deserted shores of Red Lake began to see activity again. A gang of workmen came in to clear ground and erect the first buildings, and by the end of March the first two bunkhouses and part of the cookery were ready. Arrangements were made to have shaft-sinking equipment brought in. (Ken McDougall had gone to Toronto to secure the freighting contract from Jack Hammell for this equipment. McDougall came away highly impressed with the organizers of the new enterprise but, as he remarked, they were obviously not equally impressed with him, as they decided to bring in the freight themselves.) The job was done with horses and snow tractors, the tractors being the first such to reach Red Lake. The equipment arrived on April 1, 1927, and by fall a 500′ shaft had been sunk, with lateral work underway on four levels.

Throughout 1928, underground exploration from the shaft continued, with 7500′ of underground exploration done by June. This had confirmed the existence of sufficient quantities of gold to operate a mine. There remained two major problems to be overcome: power and transportation.

The need for a source of hydro-electric power to serve the mining areas at Red Lake and to the east, coincided with existing plans to build a flood-control dam for the Winnipeg River. As early as 1926, the federal government had been investigating sites for such a dam, and Lower Ear Falls was recommended. Provisions for a generator were made in the plans for the dam, and, in June of 1928, the Ontario government announced that it would build a storage dam and power-plant at that spot.

By the fall of 1928 there were 380 men working at the Ear Falls dam. This was a major project, costing some $750,000, with the cost shared by the federal government and the governments of Ontario and Manitoba. The whole system was not complete until February of 1930. At that time the power station was completed, the Howey had built a 40.5 mile transmission line, and their plant at Red Lake was electrified.

It was obvious that a cheap transportation system was essential in order to wring a profit from the relatively low-grade gold deposits at the Howey property. The cumbersome existing system of bringing goods and supplies to the Howey was proving too expensive, and too much freight was being lost along the way.

In early 1929, Jack Hammell managed to negotiate an agreement with the Ontario government for basic improvements to the existing water route from the railway to Red Lake, to make possible the fairly easy transfer of heavy freight. The agreement specified that the Ontario government would pay 50% of the $50,000 cost of these improvements; the Howey Mine would pay 20%; and other mining properties in the area (including those at Woman Lake) would pay 30%. The improvements consisted of a series of dams to flood some small rapids and thus reduce the number of portages necessary, and of marine railways to convey goods across the remaining four portages on the Chukuni River system.

With these preliminary arrangements made, the Board of the Howey decided to begin the major developments on the property, the most important of which was the construction of a mill to process the ore. Construction and equipping of this large structure, capable of milling 500 tons per day of ore, was a complicated project as most of the necessary material had to brought to Red Lake over the primitive water/ice trail. Construction of the mill began in early 1929, and proceeded rapidly until the great stock-market crash of October, 1929.

The collapse of the mining market then left the Howey directors with an unfinished mill, an enormous list of accounts payable and an exhausted bank account. The Howey mine would have ended there, before it began, if W. S. Cherry had not offered the company a further loan of half a million dollars. In the reorganization that followed this refinancing, Hammell resigned as President of the Howey (but remained on the board). The board was able to secure, in addition to Cherry's loan, extended credit for some $300,000 worth of supplies, and development proceeded. The mill was completed in the winter months of 1930, and the mine went into production in April.

The Howey Mine made significant advances in many areas. In spite of its troubled beginnings, it rapidly became a financial success. By the end of 1930–in less than 9 months of operation–it had produced nearly half a million dollars in gold bullion. By December of 1932, the mine had paid off in full its loans from Cherry and by 1934 it became a dividend-paying operation. As a result of the mine's stringent conditions of existence (the relatively low grade of its ore, the great distance from the railroad), the Howey mine pioneered a number of innovative mining and milling techniques to cut costs. The major innovation in the mill was a "picking belt" to remove barren rock from hoisted ore before it was milled.

The shaft was eventually deepened to 2,000′, and the mill capacity increased to 1500 tons per day. The ore-body, however, became less and less rich. In 1930, the mine's first year of operation, the mined ore averaged .20 ounces of gold per ton; in 1940, its second-last year, the value was .06 ounces of gold per ton. Thus, in spite of the fact that innovation in technique and economy of scale had led to major improvements in the cost per ton of ore hoisted (from $5.46 in 1930 to $1.49 in 1940)–and in spite of the fact that the selling price of an ounce of gold had risen from $20.67 in 1930 to $38.15 in 1941–it was becoming obvious by 1941 that the end was in sight for the mine. The company planned to operate until sometime in the summer of 1942, as it seemed that, at that point, the profitable ore would be exhausted.

The demise of the Howey mine was speeded up, though, when on the morning of November 3, 1941 the central pillar supporting the shaft began to collapse. This was an ominous signal. Mining operations ceased immediately, and a salvage operation was begun to bring equipment to the surface. On the 8th, however, tremors were felt on surface. It became too dangerous then to go underground for any reason, and all operations ended. When final calculations were made, the Howey mine had produced 421,324 ounces of gold (as well as 144,253 ounces of silver, frequently produced at lode gold mines).

For most of its life, the Howey held the record among Canadian lode gold mines for the lowest operating cost per ton–and for turning a profit from the lowest grade of ore. The advances it pioneered in transportation to achieve this were even more important than those in mining and milling techniques.

The system of marine railways built by the Ontario government (along with a privately-owned marine railway built around the long portage at Ear Falls by Bob Starratt, who developed a very successful transportation company) was the main feature of a very cheap and efficient system for the transportation of freight over the 180 miles of waterways from Hudson to Red Lake. To carry freight along this novel system, a tugboat set out from Hudson, pulling behind it a "swing" of five flat-bottomed barges (scows), each carrying 15 tons of freight. When each portage was reached, a huge flatcar on a steel cable would roll down tracks on the portage and disappear underwater, beneath the scow. An engine house on the crest of the portage would then apply power, and the car and scow would move uphill, on the tracks over the portage. On the other side of the portage the car would again immerse itself in water and the scow would then float clear, in navigable water. The flatcar would then return to carry the next loaded barge across.

The author of an article in the *Canadian Mining Journal* once clocked the movement of a string of barges across the Snowshoe Portage; five loaded scows, carrying a total of 75 tons of freight, were moved across the portage in less than 35 minutes.[47] In summer, an ice-cooled barge loaded with perishable food made the trip from the railroad to Red Lake twice weekly.

Before the marine railways were installed, freight had to be unloaded at every portage and dragged by horse and cart over the portage on a rough trail that was often ankle-deep in mud. The new system was so much faster, cheaper and more efficient, that it almost immediately paid for itself. When the system was completed, the Northern Transportation Company (which at that point had a monopoly on the route between Ear Falls and Red Lake) reduced its freight rates from $80 to $45 a ton. In the first year of the marine-railway system's operation, some 2500 tons of goods were shipped to the Howey, for a saving of $87,500. The system itself had cost about $50,000.

In winter, freight was carried by diesel-powered tractors fitted with caterpillar treads, each hauling six or seven freight-laden sleighs and a caboose. The men who drove these tractor trains, as they were called, had a hazardous and uncomfortable job. In later years one such driver, Ron Tingley, described the experience. "We sat on the open tractor because a closed shelter would have taken the driver down with it when the rig went through the ice. . . . Between the cold and all the heavy clothes, I could hardly move near the end of the 8-hour shift. . . . We travelled two tractors together with a crew of 6. . . . We ate and slept in a caboose at the tail end of the train. It was a job to stay in the bunk, because of the lurching and swinging . . . We cached food along the route in the fall, enough to last until spring. Have you ever tasted an egg that's been left all winter in a root cellar?"[48]

Summer freight rates were around $20 per ton for contract shipments and $20 to $35 per ton for casual shipments. Winter freight rates were approximately three times as high. In summer

----

[47] Leslie Roberts, "Cities in the Bush, II: Red Lake Revisited," *Canadian Mining Journal,* December 1937.

[48] *Red Lake District News,* 50th Anniversary Special Issue, November 25, 1976.

it took just over 24 hours for a shipment of freight to go from the railway to Red Lake, and somewhat longer in winter.

The gold bricks from the Howey mine were shipped out to Hudson by this system, with virtually no concern for security. Often gold bricks would be left on a dock, unattended, at Gold Pines, waiting to be picked up–or unattended on the scows at Hudson overnight. A similar well-warranted trust surrounded the monthly trips of the clerk from Joe Kert's store at Red Lake, who came down to the bank at Gold Pines for money to cash the Howey Mine's payroll cheques. The clerk, Albert Kay, always travelled alone–by canoe, hitching a ride, or walking along the power line–and he was never robbed.

(The upright behavior evident in these cases, however, did not extend to the shipment of alcohol. After Prohibition ended and licensed premises were built at Red Lake, supplies of beer and liquor were regularly shipped north. In one case, a scow-load of beer arrived 17 cases short at its Red Lake destination. As the driver of the boat later remarked, "You needed a guard for beer more than for gold. Gold was no good to anyone and too heavy to carry, but beer . . . ")

The Howey Mine was responsible not just for innovations in milling and in transportation. As the first major employer of the Red Lake area, it was responsible for the first town and for the first evidences of twentieth-century civilization. After the gold rush had ended, the first doctor resident in Red Lake to provide medical services for the Howey employees–and any other residents of the area–was a Dr. Goodison. He was joined, shortly after his arrival in 1927, by his wife. (To his great surprise, he looked up one day from chopping wood and she appeared on the brow of a hill, ploughing through the snow. She later regaled her neighbours at teas and bridge parties with stories of life in their one-roomed cabin with its sloped sapling floor and red sateen curtains.)

In addition to company-constructed and company-owned bunkhouses for single men, privately-owned frame and log dwellings were rapidly constructed by men who had, or were about to go "out" and acquire, wives. Small stores, a bank, a hotel were also built.

The first school at Red Lake was opened in 1929, a one-room enterprise, housing grades one to eight in a log cabin near the lake. It was jointly funded by the Ontario government and by a monthly fee from each Howey employee. Two years later, a new school was built on a different site. This had two class-rooms, one for grades one to eight and one for high school, and teachers' living quarters were situated above.

At the same time, the organized churches were beginning to take tentative steps towards establishing themselves at Red Lake. Canon Sanderson of the Anglican Mission at Lac Seul, who had held the first church service at Red Lake under the spruce trees in June of 1926, continued to come to Red Lake four or five times a year until 1930. The United Church sent student ministers in the summers to minister to a wide territory which ranged, at times, from the CN line to Woman Lake and Red Lake. Red Lake received its first ordained ministers in the mid-thirties, with a resident United Church clergyman in 1936 and an Anglican clergyman in 1937.

Another, quite different type of establishment also came to the Howey townsite at Red Lake in the 1930s, as two red-light houses opened their doors for business. And business was apparently very brisk in the early years, as there were a large number of single men at Red Lake and virtually no single women. Both businesses offered bootlegging services, as well as the more traditional ones, games of poker were usually in operation, and one house was reputed to have the first piano at Red Lake.

In spite of these attributes, neither seems to have had the flash and dazzle associated with sister establishments in earlier mining areas like the Klondike. One of the Madams, in fact, was reputed to have been an ex-schoolteacher who lost her teaching job when she became pregnant, and she apparently contributed money to the further education of several Red Lake children. (She also reportedly offered a substantial sum of money to the building fund for the first United Church building, being planned in the late 1940s. When some of the church members demurred about accepting a contribution from so dubious a source, she apparently retorted that the United Church should be happy to accept the money, as some of it had come from its own members.)

By the late 1930s, a reasonably well-equipped town had developed at Red Lake. In an article in the *Canadian Mining Journal* of December 1937, entitled "Cities In The Bush", the journalist gave his impressions gained from a trip to Red Lake. The new hotel there would, he said, compare favourably with similar establishments in "North Bay or Sudbury . . . or any of the small cities along the Montreal-Toronto highway." And as for community life, he said, "Red Lake is any modern southern Canadian village or small town transplanted into the bush a hundred miles north of the steel: housewives with modern domestic appliances, bridge, badminton and tennis clubs, softball and hockey teams, a high school that last year had 72 students, a weekly newspaper."

(This description was not completely accurate. Red Lake was and is a typical, unplanned mining town with streets veering off in all directions and a large number of haphazardly designed and cheaply constructed houses. Its community life has always been an active one, but certain community services–like community water and sewage systems, sidewalks and paved streets–were very late in arriving, in some areas, and have yet to arrive in others.)

The development of small businesses that occurred around the Howey mine at Red Lake–along with the location of the new townsite, at the terminus of the water route north from the railway–established it as the commercial centre of the area, as other mines began to develop in the district. And this is a function the town of Red Lake has maintained, some thirty years after the last mine at Red Lake itself ceased to operate.

At the same time as the first town and the first commercial developments were growing, in order to serve the Howey's employees, the Howey was also responsible for a renewal of interest in other potential mining properties in the area. In the early 1930s, the only activity at Red Lake was at the Howey. The worldwide economic depression that followed the stock-market crash of 1929 meant that it was virtually impossible to find capital to do even the most basic development work on a mining property.

In this depressed climate, the only money and impetus for further prospecting in the Red Lake area were coming from the Howey. An article in the *Canadian Mining Journal* in 1931 pointed out that after the recession of the gold rush had left the Red Lake area high and dry, a number of syndicates composed of men from the Howey mine–and therefore indirectly funded by the Howey payroll–were beginning to do some "real" prospecting. The author of this article, after referring to the blind, blanket staking of the area, during the Red Lake gold rush, that had accomplished virtually nothing, stated his belief that there were at least half-a-dozen "first-rate prospects" in the Red Lake area, and that these would be properly prospected and developed only by "systematic prospecting on the part of men who have given special attention to the structural and geological features of a district." (The journalist's predictions turned out to be accurate on all counts.)

The solid, if unspectacular, Howey record, and the further developments that flowed from the Howey mine, were in sharp contrast to the situation in the Woman Lake area during the 1930s. The gold rush to that district (now termed the Uchi-Confederation Lake area) had stolen much of the thunder from Red Lake during the gold rush months, but the results were very different. A total of nine small gold mines were developed in the area and produced intermittently, from 1929 to 1952, with most of the production in the late 1930s. The total gold production of all the mines there was only some $8.6 million. And the only major mine ever developed in the area was a base-metal producer, owned by Stelco. The South Bay Mine produced zinc, copper and silver, between the years of 1970 and 1981.

Whereas it is true that the Howey mine sustained Red Lake for almost a decade, it took more than the impetus it could provide to bring about the real development of other mines in the area. What it took, in fact, was an increase in the price of gold. After the negative effects of the Great Depression had begun to be felt in earnest, a rapid re-evaluation of the value of gold in the international economic system occurred. On January 31, 1934, the U.S. government officially went on the gold standard, and Canada began to sell gold to the U.S. for $35 an ounce.

This prompted what has been called the "Depression Boom" in Canadian gold mining, a boom which brought in its wake an enormous amount of mining development in the Red Lake area.

During the period from 1933 to 1940, a total of seven new gold mines came into production in the Red Lake area. (Every one of these was on claims staked during the early months of the gold rush.) The first of this group of mines (and the second producing mine at Red Lake) was McKenzie Red Lake Gold Mines Limited on McKenzie Island, the largest island in Red Lake. The company went into production on March 1, 1935 with a mill treating 125 tons of ore per day. The mill tonnage was soon increased, the shaft was deepened, and a townsite sprang up next to the mine site. The town rapidly developed a number of small businesses, and for a time, in the 1940s, rivalled the town of Red Lake in commercial importance. The mine operated until March of 1965, when the supply of profitable ore became exhausted. In its lifetime it produced just over $24 million in gold bullion, and paid dividends of over $3 million to its shareholders.

In August of 1936, a smaller mine, the Red Lake Gold Shore Mines, came into production on land close to the town of Red Lake. This mine operated for only two years, in that time producing some 22,000 ounces of gold bullion valued at $738,486. In October of 1937, Gold Eagle Gold Mines went into production. This property was on McKenzie Island. It was another short-lived producer, as its ore was rapidly exhausted and it ceased production in September of 1941, after producing some 40,200 ounces of gold bullion worth about $1.4 million.

The next mine to be established at Red Lake proved to be one of the area's major producers. This was Madsen Red Lake Gold Mines, developed by Marius Madsen on a group of claims seven miles south-west of the town of Red Lake. (Most of these claims had been staked, originally, by Nehemiah Faulkenham in January, 1926.) A company-controlled townsite was laid out near the mine, with houses built by the employees and services provided by the company. Madsen began producing in August of 1938 and continued until 1972, producing almost $82 million in gold bullion during that time.

Another mine was also established at Red Lake in 1938–this was Hasaga Gold Mines, on the claims originally staked by Ray Howey for McIntyre-Porcupine Mines. This property, within the erratic boundaries of the town of Red Lake, hoisted the ore at its own shaft directly over the ore body, and then trucked the ore to its mill a mile away, which had belonged to the recently deceased Red Lake Gold Shore. Hasaga operated until 1952, producing some $8,250,000 in gold bullion during its lifetime.

The sixth mine to be developed during the "Depression Boom" was Cochenour-Willans Gold Mines, on the shores of Red Lake opposite the McKenzie Island townsite. The fact that this mine, which turned out to be one of the area's major producers, ever became a reality was a tribute to the persistence of its stakers, Ed and Bill Cochenour and Dan Willans. They had to endure thirteen years of dropped options and unfavourable reports from all manner of experts before the prospects of their claims were judged favourable enough to attract sufficient capital for development. When financing was finally secured, a mill was built in the record time of three months, and the company went into production in December of 1939. A well-planned and attractive town was built for (and by) company employees. The mine closed down in October of 1971, producing in its lifetime some $41,375,000 in gold bullion.

The seventh and last mine of this era was a short-lived one, McMarmac Red Lake, which operated from 1940 to 1944 (and was later re-opened for a year, 1948-1949). It produced a total of $2,750,000 in its brief lifetime.

The impact of all this simultaneous development was enormous. Looking at it from a contemporary point of view, the total value in gold bullion produced by those seven depression-era mines was an impressive $160,525,000. The gold production of those mines, however, was even more impressive than this figure. For during almost all the lifetime of even the longest-lived of those mines, Madsen Red Lake, the price of gold was pegged at $35 an ounce, a tenth of its current value. These isolated and–generally speaking–rudimentary communities have made a very large contribution to the economy of Canada. The settlement-in-isolation that the Red Lake area embodies has received, in its 60-year lifetime, relatively little in the way of government funding in return for the massive amounts of capital, through the production of bullion and the payment of taxes and dividends, which it has injected into the Canadian economy.

At the time of the rapid and massive developments of the 1930s, though, the atmosphere at Red Lake was one of well-warranted excitement and optimism. There was no reason why Red Lake should not become, now, a real "city in the bush". Unlike the feverish excitement of the gold rush ten years earlier, this time around the optimism was based not on hopes but on actual happenings. What was coming into Red Lake now, as one mine after another was brought into production, was not the hopeful prospector with his dog-team, but building materials, mill and underground equipment, real capital in its most solid forms.

The task of bringing into Red Lake the necessary equipment and supplies for both preliminary and final development of these mining properties was enormous. The water-route carried capacity loads from break-up to freeze-up, with a number of companies operating both freight and passenger services. In winter the tractor-drawn sleighs of freight crawled continuously up the ice-covered lakes.

Regular air service to Red Lake had continued since the gold rush, varying in frequency with the amount of activity at Red Lake. The Elliott-Fairchild air service had been disbanded in the late summer of 1926. Just before Christmas of 1926, though, Captain Harry Oaks left Patricia Airways and, financed by James Richardson of Winnipeg, founded Western Canada Airways and began flying the company's first aircraft to Red Lake and Woman Lake. (This company later became the western division of Canadian Airways, which flew large quantities of freight to Red Lake in the mid-thirties.) Patricia Airways operated a service to Red Lake in 1927, and later during the 1930s boom. Robert Starratt branched out from surface to air transportation and formed Starratt Airways, active in the 1930s. And, for a short period in 1937, British North American Airways, attempting to enter the air transport business in Canada, operated a service from Sioux Lookout to Red Lake.

For all the airlines involved in northern bush flying, the depressed period in mining activity that followed the stock-market crash had meant a drastic curtailing of flights. But as mining exploration and development took off in the Depression Boom, the airlines boomed too. The volume of air traffic to Red Lake became so great, in fact, that, in the summer months of 1936, the waters of Red Lake saw more takeoffs and landings than any airport in the world, and in the year 1937 handled an astounding 50% more air freight than any airport in the world.

The airplanes during this expansionist time were carrying an enormous variety of goods. Robert Starratt is reported to have said that "We have loaded, conveyed and discharged a complete set of ball mill liners, total weight 8700 pounds, mine hoists in sections, weighing 1400 pounds each, bulky mine cars weighing 750 pounds each, a length of heavy shafting sixteen feet long, an assay furnace complete, along with a crate of eggs, a bale of hay and a bride's trousseau to fill broken storage space and trim the ship. Canoes, live pigs, chickens and dogs are mere incidentals."[49]

(It is interesting to note the costs of flying to Red Lake at this time, compared to the costs during the gold rush. In 1940 it cost $40 return from Hudson, or $36 return from Sioux Lookout, to fly to Red Lake. The cheapest fare on the "old crates" that flew to Red Lake during the gold rush was $200 return from Hudson.)

Just as the gold mining industry boomed during the Depression, when most other industries were failing–and in fact provided the only bright spots on the Canadian financial markets during the worst years of the Depression–so, too, the war years had an effect on gold mining quite opposite to that on the rest of the Canadian economy.

In the first part of World War II, gold mining continued to benefit as the depreciation of the Canadian dollar raised the price of gold from $35.00 to $38.50 per ounce. As the war progressed, though, this beneficial effect was reversed. Maximum effort had to be put into what were termed "essential industries"

----

[49] Quoted in Rae Kiebuzinski, *Yesterday the River*, p. 274.

and gold mining, which was not deemed essential, began to suffer. When reciprocal agreements were signed between Canada and the U.S. to maintain a balance in the commodity exchange, gold bullion–which used to be very significant in maintaining that balance in Canada's favour–became much less important. At the same time the costs of equipment and supplies necessary for gold mining rose as the economy geared up, certain supplies became very difficult to procure as they were being sent to the "essential" industries instead, and the mining industry lost about half its labour force to other industries and to the armed forces.

(The gold mines had hired many foreign workers. An example of this was the so-called "International Row" at Red Lake in the late 1930s, a short street of dwellings which housed, in order, a couple from Finland; a couple from Denmark; a German family; an Irish couple; an English couple; a bachelor from Austria; Ken McDougall and his wife; an Italian with a Ukrainian wife; a French-Canadian family and a Norwegian family. In 1942, military service qualifications were changed to allow men of foreign extraction and birth to qualify for military service, and a number of employees at the Red Lake mines left for the armed forces.)

The statistics are significant in telling what happened to gold mining during World War II. Three of the seven mines at Red Lake closed: Howey, Gold Eagle and McMarmac. In 1941, just before the axe fell on the gold mines, more gold was produced in Canada than in any year, before or since: 5,345,179 ounces in all of Canada; 3,194,308 ounces in Ontario. In 1945 production had fallen to half those figures: 2,696,727 ounces in Canada, 1,625,368 ounces in Ontario.

After the war was over, though, the boom-and-bust cycle that characterizes mining, swung upwards towards the boom again. Supplies were more readily available, government regulations were less stringent, and capital was again obtainable as investors looked for new speculative areas. A great deal of staking occurred around Red Lake. A number of companies were formed to do exploration and development work on their claims, and out of this three producing mines developed.

The first of these was Starratt-Olsen Gold Mines, whose property was a mile from the existing Madsen mine. This new mine was in production for 8 years, from September of 1948 to September of 1956, and it produced around $6,000,000 in gold bullion.

The second and third mines developed during this postwar boom, Dickenson and Campbell, are the only two producing in Red Lake today. Dickenson Red Lake Mines acquired its claims from Gordon Shearn. Shearn had originally staked many of those claims in 1926 for his employer, Major Cunningham-Dunlop. A fair amount of assessment work was done on the claims over the next 18 years, revealing visible gold but not an ore-body of a size profitable to mine.

The claims changed hands several times and were eventually restaked by Shearn for himself in 1944. He then sold his claims to the newly formed Dickenson Red Lake Mines, which started production in December of 1948. The company, which has seen several corporate reorganizations, immediately became a solid, medium-sized producer and remains so today. It has interests in a number of other mining exploration and development companies, in the Red Lake area and elsewhere.

The third mine established in this period, Campbell Red Lake, has established itself as the premier gold mine in Canada, mining a grade of ore richer than any other producing gold mine. Dome Mines took an option to purchase this property during the early stages of its development in the mid-1940s, and proceeded with the option to gain control. The company began production in June of 1949 with a 300-ton-per-day mill. The mill capacity, in 1982, was increased to just over 1000 tons per day, and the end has not yet been predicted to the company's ore reserves.

The mines at Campbell and Dickenson are on adjoining properties (in fact, the two firms are mining essentially the same ore body) and the two companies constructed a joint townsite. This was unusual for the Red Lake area in which, unlike most such isolated mining areas, every major mine had its own townsite. The new town of Balmertown was a professionally-planned town, the first example in the Red Lake area of the type of instant-suburbia that has become common since in resource developments in northern Canada. Campbell Red Lake from its beginning was the pioneer among area mines in establishing employee benefits and providing recreational facilities for employees and their families–in 1982, for example, the company opened a large and relatively lavish recreational centre with Red Lake's first indoor swimming pool. It might be said that Campbell and Dickenson were well able to provide these amenities, but other major producers in the area, like Madsen, were able to pay considerable dividends, in their heyday, without seeing the need for improving rather bleak company townsites.

The development of these mines, in the 1940s, was directly responsible for two major developments in the transportation system to Red Lake. The combined systems of marine railways and bush planes had worked fairly well up to this point, but it was now becoming severely overtaxed. The Ontario government at last agreed to build a highway to Red Lake. This winding gravel road, linking Red Lake to the Trans-Canada Highway, was constructed at a cost of $2.75 million and opened in December of 1946. (It was given a hard surface in the 1960s.) This road's rough surface and hairpin curves gave the beleagured motorist the impression he was negotiating country much more rugged than the Shield, with its relatively small hills and numerous patches of muskeg.

In the next year, 1947, air traffic also became immune to the vagaries and delays of fall and spring break-up. A year-round landing strip was constructed close to the Cochenour property. Several other phenomena of civilized life appeared in Red Lake during the 1940s: the Red Cross opened a 16-bed hospital in an old Howey bunkhouse in 1946 (there had been several small hospitals opened by some of the mines before this) and regular long-distance telephone service was installed in 1947.

While the advent of long-distance telephone service represented an important breakthrough in terms of improving regular channels of communication, the initial telephone link (by radio-telephone) had been established ten years previously, in May of 1937. That event had been marked by a coast-to-coast CBC network broadcast from the lobby of the Red Lake Hotel. The CBC commentator illustrated the importance of the new link with the following story.

"Some years ago a man in Red Lake was taken very ill. The doctor was out of the area handling an emergency case. . . . The forestry radio-telegraph station had finished its regular schedule at 5 o'clock and they knew no-one would be listening at Kenora or Sioux Lookout. But the forestry operator had an idea. He changed his wave-length and got on the amateur band. His CQ [all-stations] call was answered by an amateur in Tennessee. The Tennessee man was asked to send a telegram to Sioux Lookout. Sioux Lookout located a doctor, got in touch with Red Lake, the doctor issued instructions about looking after the patient for the night, and flew in himself just as soon as it was daylight. The man's life was saved."[50]

A period of relative stability followed the developments of the late 1940s. (It was relative only, because when a community's existence is founded on a non-renewable resource, there is no real stability.) A fourth minor "boom" occurred in the late 50s and early 60s. At this time four mines were opened, all short-lived. Three of them were merely extensions of existing mines: Annco was controlled and operated by Cochenour, Wilmar also by Cochenour, and Robin Red Lake by Dickenson. The fourth mine to open then was H. G. Young Mines. Horace Young was a prominent Canadian mining engineer whose various involvements at Red Lake dated back to 1927, when Jack Hammell recruited him as chief engineer for the Howey. The mine he founded, however, had a far less successful career than the man himself, as it was in production for only two and a half years.

It was in the 1950s that the scattered townsites of the isolated Red Lake area attained an air of something like permanence, and life there lost its pioneering flavour. (This was not an unmixed blessing, as some of the early, very strong community spirit was inevitably lost.) Growing up in Red Lake in the 1950s was much like growing up in a small town anywhere in Cana-

[50] *Red Lake District News*, May 31, 1937.

da, with the big city–Winnipeg–within a day's drive. Even television came to Red Lake (albeit in a very fuzzy form) in the late 50s. The first generation of children born in Red Lake grew to maturity in the 1940s and 1950s. Some stayed in Red Lake and worked for one of the mining companies (there was very little employment, though, for women, so that unless a girl married young, she left Red Lake); some drifted to other mining communities; some went to Winnipeg or Thunder Bay for training or for jobs; a few went "out" to university.

Even during this relatively stable and prosperous time, though, most of the mines were becoming caught in an increasingly tight economic squeeze which grew tighter in the 1960s. The federal government's Emergency Gold Mining Assistance Act, which operated from 1952 to 1976, was of some help–but not enough. In gold mining, as in few other industries, the product of the industry had been fixed in price since the 1930s, while overhead costs rose steadily. (This was particularly true of the cost of labour, even though the wage scales of the gold mines–except at the top management levels–were appreciably lower than those of other industries.)

Another factor in the cost-price squeeze was one which affects all mines sooner or later: the good ore began to run out, with the remaining ore, more and more difficult and expensive to mine. Many of the people, who had come to Red Lake in the boom days and raised their families, moved away. McKenzie Red Lake stopped production in 1965, after 30 years; Cochenour in 1971, after 32 years; Madsen Red Lake in 1972, after 34 years.

A bright spot in the life of the community of Red Lake occurred in the late 1960s when a major iron ore production facility was built at Bruce Lake, about 30 miles down Highway 105 from Red Lake. The iron-ore pelletizing operation, owned by Stelco, was called the Griffith Mine. It cost some $62 million to build and provided an enormous boost to the economy of Red Lake–in the stages of construction, and later, as a source of work for former employees of the recently-closed mines, or for the second generation of Red Lake families.

The Griffith Mine did not build a townsite for employees, and as a result the housing industries in both Red Lake and Ear

Falls, to the south, were given a boost. (And much of the housing at the now-closed mines was thus prevented from the sorry fate of abandonment and decay.)

With the closing of the mines at Madsen and Cochenour in the early 1970s, only Campbell and Dickenson remained in the area that had once supported half a dozen major operations simultaneously. The production of gold of the two remaining operations, though, was very significant. In 1974, with just two mines operating, the gold production of the Red Lake area, for the first time, surpassed that of the Porcupine district. In 1978, Campbell Red Lake produced the hefty amount of 5,708.9 kilograms (201,373 ounces) of gold, more than a quarter of all the gold bullion produced by all the gold mines in Ontario, and more than a tenth of the total amount of gold produced in Canada. The importance of the town (actually, the Improvement District) of Balmertown, the joint townsite of Campbell and Dickenson, increased. Red Lake remained the main commercial centre, but Balmertown increasingly became the location of organized recreational facilities.

When the price of gold skyrocketed in the late 1970s and peaked in 1980, there were immediate speculations that gold-mining would have another great boom. There was an immediate and dramatic increase in the number of claims staked in the Red Lake area, and in the amount of exploration there. All available land was staked, and many properties became optioned to various corporations.

As with most other such bursts of optimism throughout Red Lake's history, though, this one materialized only in a diluted form. Campbell Red Lake expanded its mining and milling operation, and Dickenson announced a program (not yet completed) to increase its milling and hoisting capacity. And Wilanour Resources, controlled by the large Camflo corporation, began operations at two recently-acquired properties in the Red Lake area, one just to the south-west of the town of Red Lake and the other on the old Cochenour property. Both these operations, however, were dependent on gold remaining at a very high price. When it dropped to $325 an ounce in early 1982, they were discontinued.

While Campbell Red Lake has never undergone any financial crises (and, unofficially, people at Campbell talk about 100 years more of mining operations), the operation at Dickenson recently weathered a fairly severe financial storm. A period of retrenchment, followed by an agreement with another mining corporation trading a substantial interest in the gold mine for a substantial infusion of cash, seems to have put the company firmly on its feet again.

A visit to the Red Lake area today leaves one with mixed impressions. The town of Red Lake is a lively and not unattractive place in mid-summer. The constant noise of small aircraft taking off and landing from the waters of Howey Bay rings in the hot summer air. Most of the stores with their boom-town fronts are well-maintained, and substantial new homes sit cheek-by-jowl with older frame structures on the rambling unpaved streets. The town of Balmertown is clean and prosperous, and so is the community of Cochenour, even though it has been without an operating mine since 1971.

But, eight miles to the southwest of Red Lake along Highway 618, the community of Starratt-Olsen is a living example of the end that most mining communities come to. At this townsite the mine buildings have been vandalized, piles of scrap lumber and old cars dot the once-neat front yards. Most of the houses that were not demolished by order of the fire department are still inhabited–but only a few of these exhibit any pride of ownership. The forest (or the "bush", as it is called) is closing in again.

Meanwhile, though, the rest of the area carries on in a holding pattern–waiting for the price of gold to rise again, for the steel mills of southern Ontario to, once again, need all the output of the Griffith Mine, for more tourists to arrive. There will not be another gold rush like the one of 1926, for, in the 1980s, that kind of unbounded optimism no longer exists. On the other hand, there is always the possibility that, if the price of gold rises sharply, new mines could yet open in the area, for much of it has not been explored in complete technical detail. Red Lake faces an uncertain future, but, in this respect, it is far from alone among Canadian communities.

Postscript

Since this manuscript was completed, there has been a resurgence of interest in gold mining in Canada as the price of gold has increased steadily, if slowly, to the $400.00 range. New gold mines have opened across the country, from New Brunswick to British Columbia.

The most startling and dramatic Canadian mining story of the 1980s has been the development of the Hemlo gold field near Marathon, on the north shore of Lake Superior. A 1980s version of the traditional gold rush followed the initial discoveries; contemporary features included, not only the expected high-tech development, but a lengthy and acrimonious lawsuit between International Corona Resources and LAC Minerals, owners of two of the very rich mines in the area.

Three major mines have been opened at Hemlo, and their combined production in 1985 was a full three of the 86 metric tonnes of gold produced by all mines in Canada. It is predicted that by the time the three mines reach full production capacity in 1989, the annual production from the Hemlo gold field should be approximately 25 tonnes. A very high grade of ore is being mined in these large operations, and Noranda's Hemlo mine, for example, has a low operating cost in the range of $100 per ounce.

There have been changes in the Red Lake area as well. The Griffith Mine at Bruce Lake closed officially on April 1, 1986. Developments in the gold fields at Red Lake, though, have been all positive. Dickenson Mines completed their proposed expansion of production facilities. And exploration and feasibility work have been completed, at the time of writing, for a new mine. This is the McFindlay Mine near the East Bay of Red Lake, close to the Campbell Red Lake site; a final decision on its future is expected shortly.

And at the corporate level, a major amalgamation has occurred: Campbell Red Lake Gold Mines, Dome Mines and Placer Development Corporation have joined to become Placer Dome Incorporated, now the biggest gold producer in the western world outside South Africa.

As I write this manuscript, plans are in the final stage for the first Red Lake District High School reunion in July, 1987. The official reason for the reunion is the 35th anniversary of the opening of the high school, in 1952. The impending reunion has caught the imagination of the alumni of this small school who have spread across Canada, and an astonishing 600 people are expected to converge on the town. They are coming for a reason; Red Lake never became a "city in the bush" but it was a real community whose culture formed all of us who grew up there.

Kitchener, Ontario
June, 1987

Bibliography

Barr, Elinor, and Betty Dyck. *Ignace: A Saga of the Shield.* Winnipeg: Prairie Publishing, 1979.

Berton, Pierre. *Klondike.* Revised edition. Toronto: McClelland and Stewart, 1972.

Bishop, Charles A. *The Northern Ojibwa and the Fur Trade.* Toronto: Holt, Rinehart and Winston, 1974.

Bishop, Olga. *Bibliography of Ontario History.* 2 volumes. Ontario Government: Ontario Historical Studies Series. Toronto: University of Toronto Press, 1980.

Blythe, Jack. *The Canadian Social Inheritance.* Toronto: Copp Clark, 1972.

Botterill, B., and V. Gouriluk, Y. Karim, E. Milliken. *Profile of Red Lake.* Red Lake: Township of Red Lake, 1974.

Bradwin, Edmund W. *The Bunkhouse Man: A Study of Work and Pay in the Camps of Canada, 1903-1914.* Toronto: University of Toronto Press, 1972.

Brown, L. Carson. "The Red Lake Gold Field." *Canadian Geographical Journal,* April 1965.

Bruce, E. L. "Geology of the Basin of Red Lake, District of Patricia." *Annual Report of the Ontario Department of Mines.* Toronto: King's Printer, 1925. Vol. 33, part 4 (1924):12-39.

Bruce, E. L., and J. E. Hawley. "Geology of the Basin of Red Lake, District of Kenora (Patricia Portion)." *Annual Report of the Ontario Department of Mines.* Toronto: King's Printer, 1928. Vol. 36, part 3 (1927):1-72.

Bryce, George. *The Remarkable History of the Hudson's Bay Company*. Toronto: William Briggs, 1900.

Bucksar, R. G. *Bibliography of Socio-Economic Development of Northern Ontario*. Ottawa: Canadian Research Centre for Anthropology, 1968.

Campbell, Marjorie W. *The North-West Company*. Toronto: Macmillan, 1957.

Canada Department of Energy, Mines and Resources. Mineral Resources Branch. *Canadian Minerals Yearbook*. Annual publication. Ottawa, 1956-1978.

Canada Department of Mines. *The Canadian Mineral Industry*. Annual publication. Ottawa: 1934-1956.

Canada Department of Mines. *Gold in Canada*. Ottawa: 1935.

Canadian Annual Review of Public Affairs. Ottawa: 1924-25, 1925-26, 1926-27.

Canadian Forum. Vols. 5 (1924-25), 6 (1925-26), 7 (1926-27).

Canadian Institute of International Affairs. *The Canadian Economy and Its Problems*. Eds. H. A. Innis, A. F. W. Plumptre. Toronto: Canadian Institute of International Affairs, 1934.

Canadian Institute of Mining and Metallurgy, Monthly Bulletin. 1925 to present.

Canadian Mining Journal. 1931 to present.

Canadian Red Cross Society. *Ontario News Bulletin*, June 1972.

Chisholm, E. O. "Geology of Balmer Township." *Annual Report of the Ontario Department of Mines*. Vol. 60, part 10(1951). Toronto: 1954.

Christian Science Monitor, 1 Feb.-31 Mar. 1926.

Cook, Ramsay. *Canada: A Modern Study.* Toronto: Clarke Irwin, 1963; rev. ed. 1977.

Davidson, G.C. *The North West Company.* New York: Russell and Russell, 1967.

District News (Red Lake). 50th Anniversary Special Issue, 25 Nov. 1976.

District News (Red Lake). 50th Anniversary Issue, 16 July 1980.

Dryden Observer, "Red Lake Area and Ear Falls Supplement." 27 Nov. 1969.

Dryden Observer, "Patricia Development Region: Industrial Progress in a Land of Promise." 12 Nov. 1970.

Ellis, Frank H. *Canada's Flying Heritage.* Rev. ed. Toronto: University of Toronto Press, 1962.

Fort William Times-Journal, 15 March 1926.

Gelman, Robert M. "Economic Development of the Mining Industries of the Northern Frontier Under a Planned and a Free Economy: The Soviet Union and Canada." Unpublished Ph.D. dissertation, Catholic University of America, Washington, 1964.

Glazebrook, G. P. de T. *Life in Ontario: A Social History.* Toronto: University of Toronto Press, 1968.

Hanson, Conrad H. *Gems of the North.* Kirkland Lake: privately published, 1979.

"Heritage Conservation, West Patricia. Part I: Summary and Recommendations." Unpublished manuscript. Ontario Ministry of Culture and Recreation, 1981.

Hewitt, W. H. "Transportation from the Steel to Red Lake." *Canadian Mining Journal* 55(1934):468-69.

Heidenrich, Conrad E., and Arthur J. Ray. *The Early Fur Trades: A Study in Cultural Interaction.* Toronto: McClelland and Stewart, 1976.

Holden, James E. "Red Lake Goes Modern." *The Beaver,* Dec. 1933.

Horwood, H. C. "Geology and Mineral Deposits of the Red Lake Area." *Annual Report of the Ontario Department of Mines.* Vol. 49, part 2 (1940). Toronto: 1945.

Innis, H. A. "An Introduction to the Economic History of Ontario from Outpost to Empire." *Essays in Canadian Economic History.* Toronto: University of Toronto Press, 1956.

Innis, H. A. "Settlement and the Mining Frontier." *Canadian Frontiers of Settlement.* Vol. 10. Toronto: Macmillan, 1936, pp. 167-424.

Jones, Laurence F., and George Lonn. *Historical Highlights of Canadian Mining.* Toronto: Pitt Publishing, 1973.

Kerr, D. G. G. *Historical Atlas of Canada.* Rev., 3rd ed. Toronto: Nelson, 1975.

Kiebuzinski, Rae. *Yesterday, the River: A History of the Ear Falls District.* Ear Falls: Township of Ear Falls History Committee, no date [1970's].

Kruchak, J. et al. "The Aging of the Red Lake District." Unpublished manuscript. Red Lake Office, Ministry of Natural Resources.

LeBourdais, D. M. *Metals and Men: The Story of Canadian Mining.* Toronto: McClelland and Stewart, 1957.

Lock, Alfred George. *Gold: Its Occurrence and Extraction.* London: E. and F.N. Spon, 1882.

Lougheed Associates, William. *The Gold-Mining Community: A Study of the Problems of Economic Growth.* Toronto: William Lougheed Associates, Consulting Economists, 1958.

Lucas, R. A. *Minetown, Milltown, Railtown: Life in Canadian Communities of Single Industry.* Toronto: University of Toronto Press, 1971.

Long, John. *Voyages and Travels of an Indian Interpreter and Trader.* London: privately published, 1791. Reprinted, New York: Johnson Reprint Corp., 1968.

MacDougall, J. B. *Two Thousand Miles of Gold: From Val d'Or to Yellowknife.* Toronto: McClelland and Stewart, 1946.

McDougall, K. H. "Early Red Lake Days." Address given to the Gyro International Convention, 1952. Unpublished, held in the Public Archives of Ontario.

McEvoy, Bernard L. *Canada's Buried Treasures: The Story of Minerals and their Discovery.* Toronto: McClelland and Stewart, 1968.

McInnis, Edgar. *Canada: A Political and Social History.* New York: Rinehart, 1947.

Madsen, Marius. *Shipwreck and Struggle.* Toronto: Pitt Publishing, 1963.

Mamen, C. "Stelco's Griffith Mine: An Iron Tonic for Red Lake." *Canadian Mining Journal,* Nov. 1968, pp. 48-52.

Manitoba Free Press, 1 July 1925-31 Dec. 1926.

"Man-Made Heritage Studies and the West Patricia Land Use Plan: A Status Report." Unpublished manuscript, Ontario Ministry of Culture and Recreation, 1979.

Mars, Mabelle Duncan. Mars papers, unpublished. Red Lake Office, Ministry of Natural Resources.

Mohide, T. P. *Gold.* Ontario Mineral Policy Background Paper No. 12. Ontario Ministry of Natural Resources, 1981.

Molson, K. M. *Pioneering in Canadian Air Transport.* Winnipeg: J. Richardson, 1974.

Moore, E. S. *Canada's Mineral Resources.* Toronto: Irwin and Gordon, 1929.

Morell, E. W. P. *The Gold Rushes.* London: Adam and Charles Black, 1940; rpt. 1968.

Morse, Eric W. *Fur Trade Canoe Routes of Canada: Then and Now.* Toronto: University of Toronto Press, 1979.

New York Times, 1 July 1925-31 Dec. 1926.

Northern Miner, various issues.

Ontario Department of Tourism and Information, Historical Branch, Press Release: "Plaque to Commemorate the Early Fur Trade in Red Lake." 1 Sept. 1965. Public Archives of Ontario.

Ontario Department of Travel and Publicity, Historical Branch, Press Release: "Plaque to Commemorate Founding of Red Lake Mining District." 20 Sept. 1962. Public Archives of Ontario.

"The Ontario Division in Red Lake District." *The Canadian Red Cross,* Sept. 1926.

Ontario Ministry of Culture and Recreation. "At the End of the Rainbow: A Study of the Gold-Mining Region at Red Lake." Unpublished manuscript. 1977.

Ontario Ministry of Natural Resources. *Ontario Mining Statistics: a Preliminary Compendium*. Toronto, 1980.

Pain, S. A. *Three Miles of Gold: The Story of Kirkland Lake*. Toronto: Ryerson, 1960.

Pain, S. A. *The Way North: Men, Mines and Minerals*. Toronto: Ryerson, 1964.

Parrott, D. F. *The Red Lake Gold Rush*. 4th edition. Red Lake: privately published, 1974.

"Red Lake: An Experiment in Housing for Indians." *Ontario Housing* 10(Feb. 1964):11-15.

"The Red Lake Gold Area." *Canadian Mining Journal* 52(1931):589-92.

Red Lake News, 31 May 1937; later this newspaper was called *Red Lake District News*, and *The District News*.

"Red Lake Romance." *Gold, Magazine of Canada's North*, Jan. 1940, pp. 7-31.

Rich, E. E. *The Fur Trade and the Northwest to 1857*. Toronto: McClelland and Stewart, 1967.

Roberts, Leslie. "Cities in the Bush, II: Red Lake Revisited." *Canadian Mining Journal*, Dec. 1937.

Robinson, H. M. *The Great Fur Land, or, Sketches of Life in the Hudson's Bay Territory*. New York: Putnam, 1879.

Robinson, Ira M. *New Industrial Towns on Canada's Resource Frontiers*. University of Chicago Program of Education and Research in Planning, Research Paper No. 4. Chicago: University of Chicago Press, 1962.

Rowe, R. "Brief Historical Sketch of the Howey Mine." *Canadian Mining Journal* 55(1934):434-37.

Spencer, Loraine, and Susan Holland. *Northern Ontario: A Bibliography*. Toronto: University of Toronto Press, 1968.

Shearn, Gordon. "Early History of Red Lake." *Precambrian*, April 1956, pp. 9-14.

Times (London), 1-31 March 1926.

Toronto Star, 1 July 1925-31 Dec. 1926.

Townsley, B. F. *Mine-Finders: The History and Romance of Canadian Mineral Discoveries*. Toronto: Saturday Night Press, 1935.

Tucker, Albert. *Steam Into Wilderness, 1902-1962*. Toronto: Fitzhenry and Whiteside, 1978.

Vinet, Robert Gary. "An Analysis of the Educational Effort of a Single-Enterprise Community: Red Lake." Unpublished M. Ed. thesis, University of Manitoba, 1970.

Voorhis, Ernest. *Historic Forts and Trading Posts of the French Regime and of the English Fur Trading Companies*. Ottawa: Department of the Interior, 1930.

Wallace, W. Stewart, ed. *Documents Relating to the North-West Company*. New York: Greenwood Press, 1968.

Watson, E. H. A. *History, Ontario Red Cross, 1914-1946*. Toronto: Red Cross Ontario Division Headquarters, 1946.

Index

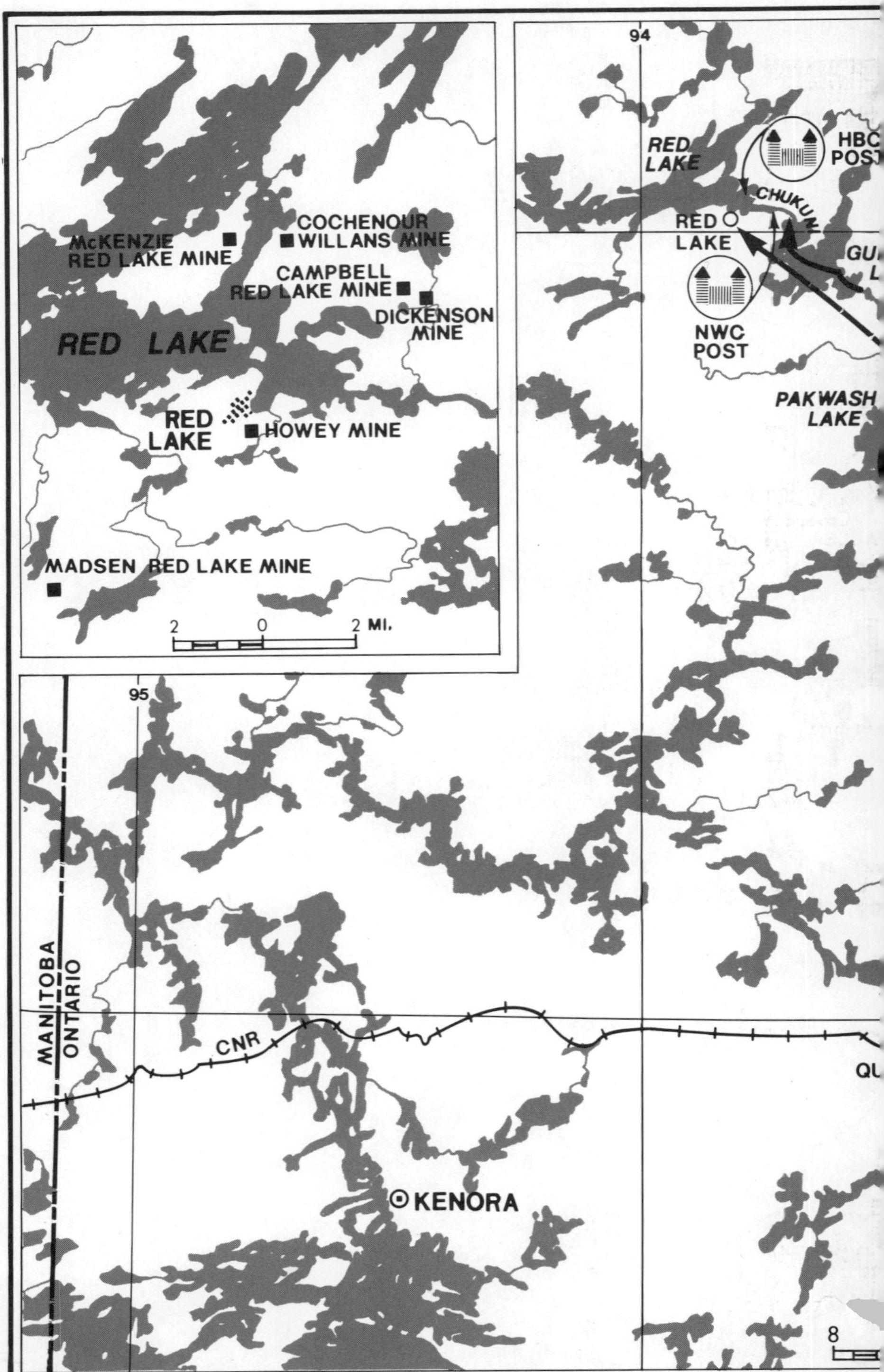

McKENZIE RED LAKE MINE
COCHENOUR WILLANS MINE
CAMPBELL RED LAKE MINE
DICKENSON MINE
RED LAKE
RED LAKE
HOWEY MINE
MADSEN RED LAKE MINE
2 0 2 MI.
94
RED LAKE
HBC POST
CHUKUNI
RED LAKE
NWC POST
PAKWASH LAKE
95
MANITOBA
ONTARIO
CNR
KENORA
8